Ecosystem services provided by agricultural areas

Evaluation and characterisation approaches

Anaïs Tibi and Olivier Therond

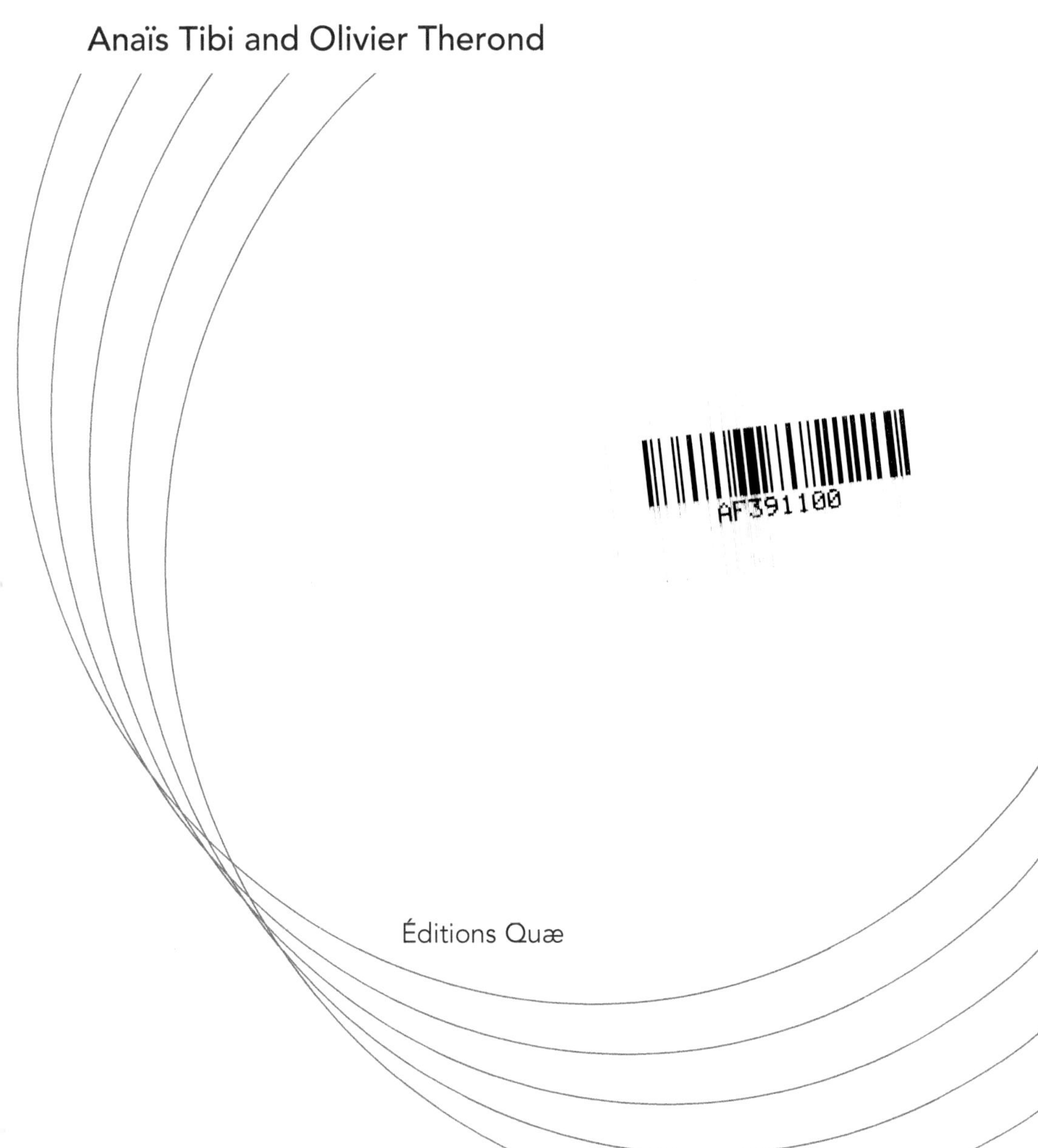

Éditions Quæ

The publication of this work received funding from the Directorate for Collective Scientific Assessment, Foresight and Advanced Studies (DEPE) of the French National Institute for Agriculture, Food and the Environment (INRAE, former Inra).

This book is the result of an advanced study commissioned by the French Ministry in charge of the Environment. The study was drafted by a committee of scientific experts without prior approval requirements from either the sponsors or INRAE.
This work is the sole responsibility of its authors.

The documents pertaining to this study are available on the INRAE websites (https://www.inrae.fr/).

The authors of the scientific report, which is the basis of this work, are all the members of the study workgroup listed at the end of the book. Unless otherwise specified, the figures and tables were produced by the authors. The websites mentioned in this book were accessed in June and July 2023.

To quote this book:

Tibi A., Therond O., 2024. *Ecosystem services provided by agricultural areas. Evaluation and characterisation approaches*, Versailles, Éditions Quæ, 172 p.

Translated by Laura Sayre.

Content

Preface

By responding to the request for evaluation of agroecosystem services in France, as part of the national EFESE program supported by the Ministry in charge of the Environment, the work coordinated by the Directorate of Collective Scientific Assessment, Foresight and Advanced Studies, of which this work presents a synthesis, has taken up several challenges.

The first challenge is to have, for more than two years, brought together around forty scientific experts from various disciplinary fields and institutional origins, and mobilised Inra's (now INRAE since 2020) skills in modeling agricultural ecosystems, data engineering and cartography. A summary of a report of nearly 1,000 pages, this work presents the conceptual and methodological advances, and the main results obtained by this multidisciplinary experts committee.

The second challenge is conceptual. The transposition of the notion of ecosystem service to the case of agricultural ecosystems, which are highly managed or even constructed, is not obvious. It therefore required developing an important and original conceptualisation, and making choices in a scientific field where debate is intense. The purposes of agriculture and the plural nature of agricultural practices have led the expert committee to differentiate goods and services, and to distinguish the practices those which build the ecosystem - the installation of planned biodiversity - from those which relate to the provision of exogenous inputs such as water, fertilisers and phytosanitary products and which regulate the potential for ecosystem services. In the same vein, proposals were made to clarify the oppositions between services, disservices, positive and negative externalities of agriculture.

A third challenge lies in the specification of services and the choice of biophysical and economic evaluation methods, the two essential dimensions of the notion of ecosystem service. This specification and this assessment — enriched by previous conceptual reflection, including particular attention to the links between services, benefits and beneficiaries, and an overview of the specificities of French agricultural ecosystems — required profound adaptations of the international typology of ecosystem services (Common International Classification of Ecosystem Services)[1] and a rich and profitable review of evaluation methods. A total of 14 services, which offer good coverage of the categories "regulatory services", "goods" and "cultural services", were studied. The use of the finest possible spatial resolution (down to the field), databases on soils, climate and cropping systems, and crop and meadow simulation models, results in an assessment based on both precise and complete throughout the national territory.

---

1. https://cices.eu/

A particular strength of this evaluation, beyond the information it provided to inform public decision-making, is to have fed back into the initial conceptual reflection around the relationship "agricultural practices – biodiversity – service – benefit". For example, a quantification at the national scale of the share of agricultural production enabled by input ecosystem services and that enabled by the provision of exogenous inputs was carried out. Likewise, this work offers an enlightening comparison of the maps of services provided by agricultural ecosystems and the negative impacts of agriculture relating to similar criteria, such as the regulation of water quality by immobilisation of mineral nitrogen (service) and the amount of leached nitrogen (negative impact). Similarly, the economic assessment prompted critical reflection from the authors on the conditions for applying the methods and the need for a solid biophysical evaluation upstream.

The consideration of service bundles, which is crucial for rethinking the management of agricultural ecosystems, emerged as an additional challenge. This work has given a unique place to the analysis of interactions between services, in which the central role of cultural sequences appears, and which makes it possible to identify major management levers.

The perspectives opened by this work are rich on the conceptual, methodological and cognitive levels. Understanding the role of livestock and the management methods of the agroecosystem in the provision of ecosystem services are of course central to these perspectives. Likewise, the relationship "planned biodiversity – associated biodiversity" and the key role of biodiversity in the provision of services must still be deepened and explained. The results of these investigations, which call for a renewal of approaches, are highly anticipated as the potential impacts are significant.

By meeting all the challenges mentioned above, the expert committee not only responded to the request of the Ministry in charge of the Environment, but it also shared its considerable work and its achievements with the French community gathered in the Inra unifying program on agricultural ecosystem services, which supported the project with great interest. Even more, these experts contributed to advancing the thinking of researchers who invest in the fundamental area of the links between agriculture, biodiversity and the concept of ecosystem service, at the interface between science and society. Before inviting the reader to delve into this work, we would like to thank the 71 members of the working group who contributed to this major project, providing a solid scientific foundation and paving the way for numerous future works on characterisation and assessment of ecosystem services.

Guy Richard, Françoise Lescourret
First Inra research programme on agricultural
and forest ecosystem services - EcoServ (2013-2019)

Main acronyms used

C: carbon

CH_4: methane

CICES: Common International Classification for Ecosystem Services

CO_2: carbon dioxyde

DOC: dissolved organic carbon

DEPE: directorate for Collective scientific assessment, Foresight and advanced Studies

EFESE: *Évaluation française des écosystèmes et des services écosystémiques* (French assessment of ecosystems and ecosystem services)

CSA: collective scientific assessment

GHG: greenhouse gas

LPIS: Land Parcel Identification System

MAES: Mapping and Assessment of Ecosystem and their Services

N: nitrogen

N_2: dinitrogen

N_2O: nitrous oxide

NO_3^-: nitrate

P: phosphorus

SAR: small agricultural region

ES: ecosystem service

PCU: pedo-climatic unit

Introduction

Context and scope of the question asked to Inra

Although the idea of "services provided by nature" appeared in the second half of the 19th century, the term "ecosystem services" first appeared in 1970, in a report known as the *Study of Critical Environmental Problems* (SCEP)[2]. Sponsored by the Massachusetts Institute of Technology, the SCEP was the first large-scale study seeking to draw attention to the global environmental impacts of human activities. In the early 2000s, the "ecosystem services" concept gained further recognition with the Millennium Ecosystem Assessment (MEA, 2005)[3], undertaken at the order of the UN Secretary General in 2000. The MEA sought to provide a scientific evaluation of the current and potential future threats to the ecosystems on which human life and wellbeing depend.

Following the MEA, in 2011 the European Union adopted a strategy intended to halt biodiversity loss by 2020. The "EU 2020 Biodiversity Strategy" is organised into six main targets, the second of which calls upon the EU Member States, with the support of the European Commission, to engage in the mapping and assessment of ecosystem conditions and ecosystem services for their respective national territories. In 2013, a dedicated working group was created – the Mapping and Assessment of Ecosystems and their Services (MAES) – the first responsibility of which was to develop of an analytical framework for Member States to employ for these assessments, in order to assure their completion in a coherent and uniform fashion.

Beginning in 2009, the French government has been engaged in advancing the MEA goals at the national level. The EFESE program (for *Evaluation Française des Ecosystèmes et des Service Ecosystémiques*)[4], launched in 2012 by the Ministry in charge of the Environment, seeks to create tools for ecosystem services assessment, for a range of different types of ecosystems, in order to improve public awareness of the value of biodiversity and to inform national and local processes for planning and development. Another objective of the EFESE program is to establish values for biodiversity within national accounting systems. The scope of this program includes all terrestrial and marine ecosystems for mainland France and its overseas territories, divided into six major ecosystem types, each of which is the focus of a thematic study:

---

2. (1970) The Williamstown Study of Critical Environmental Problems, Bulletin of the Atomic Scientists, 26:8, 24-30, DOI: 10.1080/00963402.1970.11457855. https://mitpress.mit.edu/books/mans-impact-global-environment
3. Millennium Ecosystem Assessment, 2005. *Ecosystems and Human Well-being: Synthesis*. Island Press, Washington, DC. http://www.millenniumassessment.org/en/index.html
4. https://www.ecologique-solidaire.gouv.fr/levaluation-francaise-des-ecosystemes-et-des-services-ecosystemiques

forested ecosystems; agricultural ecosystems; urban ecosystems; wetlands; coastal and marine areas; and high-mountain and rocky areas.

In this context, in early 2014, the French Ministry in charge of the Environment requested Inra (which will become InraE in 2020) to complete the "agricultural ecosystems" portion of the EFESE program (hereinafter referred to as EFESE-AE). As this request is part of a program intended to support public decision-making, Inra has entrusted the carrying out of this work to its Directorate for Collective Scientific Assessment, Foresight and Advanced Studies (DEPE). The unifying research program led by Inra since 2013 on the services provided by agricultural and forest ecosystems has joined forces with MTES to support this operation.

The goal of EFESE-AE was to describe the underlying mechanisms for a range of ecosystem services based on available scientific knowledge, and then to propose methods for biophysical and economic assessment of these services at the national level, at the most detailed scale possible. It was also a matter of identifying issues that were poorly understood and for which additional work seemed a priority. Finally, this work was to contribute to building a sustainable information system for the evaluation of agricultural ecosystems and associated ecosystem services, managed by Inra, and made available to the scientific community. Accordingly, all the assessment methods proposed and implemented by the expert group during the time available for the study were designed to be fully traceable and reproducible.

 ## Organisation of work conducted by Inra

Among the range of activities coordinated by the DEPE to inform public policies and debate, this work was organised in the form of a study, in compliance with guidelines established by Inra for the conduct of Collective Scientific Assessments (see Box 1). Approximately forty scientific experts and other scientific contributors from a variety of public institutions and with complementary disciplinary expertise (ecology, agronomy, hydrology, animal science, economics, etc.) were called upon to complete the work. Expertise in data management, a key component of EFESE-AE, was provided primarily by Inra. The working group was led by two scientific leads who directed the expert group from a scientific perspective, and a project leader responsible for the overall project management. A list of working group's members may be found on the final page of this book.

First, the experts assembled and analysed the relevant international scientific literature in order to establish an analytical framework for the specification and assessment of ecosystem services from agricultural ecosystems; determine the list of services to address; and propose indicators to assess these services. Second, these indicators were quantified using data for France; and results were analysed and interpreted.

All the information produced is included in the extended scientific report[5], written by the experts and delivered in 2017. Then, a condensed report[6], based on the extended scientific report, was written with the coordination of the DEPE between May and October 2017. The present book is adapted from the condensed report.

This document is intended for a non-specialist public and seeks to provide an overall view of the study's findings. It may be considered as a reading guide to the extended report, which is the primary deliverable for the study. All the deliverables are available *via* the Inra web site[7]. NB: This book does not include the bibliographic references reviewed by the expert group, and which support the content presented here. An exhaustive list of these references may be found in the extended report.

This book presents the results of the work carried out by the working group between November 2014 and March 2017. Chapter 1 presents the analytical framework developed specifically for the study of services provided by agricultural ecosystems, an ecosystem type characterised by a high level of anthropisation. Chapter 2 addresses so-called "input" ecosystem services provided to farmers in their role as managers of these ecosystems. A first estimate of the contribution of input services to agricultural production is proposed in chapter 3. Chapter 4 presents regulating services provided by agricultural ecosystems to the society as a whole. It also discusses possible definitions of cultural services. Chapter 5 presents available methods of economic assessment and the challenges inherent in their application to the ecosystem services concept. Chapter 6 presents an integrated analysis of the various services covered in the preceding chapters, and suggests avenues for thinking about the management of ecosystem services. A concluding section presents, in a transversal fashion, the major directions for future research suggested by this work.

---

5. https://dx.doi.org/10.15454/prmv-wc85
6. https://dx.doi.org/10.15454/1h4z-tq90
7. https://www.inrae.fr/en/news/assessing-services-provided-agricultural-ecosystems-improve-their-management

Box 1. Scientific assessment to inform public decision-making

Created in 2010, Inra's (now INRAE since 2020) Directorate for Collective Scientific Assessment, Foresight and Advanced Studies (DEPE) has the mission of informing public decision-making on complex societal issues and, at the same time, promoting reflection on the Institute on its own scientific orientations. Through the three types of exercises that it most often carries out at the request of public authorities, the DEPE is at the interface between political decision-makers, stakeholders, scientific institutions and experts. The "agricultural ecosystems" component of the EFESE program was carried out by Inra by adopting the method and principles established by its DEPE for the conduct of Collective Scientific Assessment (CSA), of which the Advanced studies are derivatives.

The institutional activity known as CSA, undertaken by Inra since 2002 and governed by a French national charter signed in 2011, is defined as a process of knowledge gathering and analysis covering a wide range of disciplinary fields relevant to public policy. It identifies existing scientific knowledge, points of uncertainty, notable areas of scientific debate, and questions requiring further research. The CSA is not intended to provide specific recommendations or practical answers to the questions confronting decision makers. The DEPE also coordinates and carries out Advanced Studies, activities that extend a CSA project by assembly and treatment of available data (statistical analyses, calculations, simulations using existing proven computer models, meta-analyses, etc.) based on published scientific work. All such exercises lead to the production of an extended scientific report written by the experts, a condensed report and a summary report both written by the DEPE.

CSA and Advanced Studies are conducted in accordance with guidelines designed to guarantee the integrity and robustness of project outputs.[8] Core principles include: competence and diversity of contributing experts (identified by Inra from their publications); impartiality (experts are required to file conflict of interest declarations, which are reviewed by Inra's Ethics Oversight Committee); transparency with respect to the methodologies adopted; and traceability of the actions taken and means employed for the project.

---

8. https://dx.doi.org/10.17180/6x6d-wn26

1. A framework for assessing ecosystem services from agricultural ecosystems

The concept of ecosystem services (ES) has gained broad currency within both the scientific community and the public policy arena, particularly in the years since the publication of the Millennium Ecosystem Assessment (MEA, 2005). A considerable amount of research and other work undertaken since the late 2000s has sought to describe and specify interactions between nature and human wellbeing. They aim to clarify the concept of ES so as to operationalise its use for decision-making and policy making purposes. Nevertheless, multiple conceptual frameworks for ES exist; these frameworks are continually evolving and subject to ongoing debate. Persistent ambiguities with regard to both the biophysical and the socioeconomic factors involved make it difficult to compare research findings across different contexts.

The ecosystem services concept as applied to the functioning of agricultural ecosystems

Agricultural ecosystems: human-impacted ecosystems managed for the production of biomass

From the perspective of both ecological and agronomic sciences, the agroecosystem is made up of two interacting systems: an ecological (or biophysical) system and a socioeconomic system. In this book, the term "agricultural ecosystem" refers to the ecological system; or in other words, the soil and its vegetation cover, associated semi-natural features (hedgerows, isolated trees, wet areas, field margins, etc.) and animals living in or passing through the field (livestock on pasture, wild animal biodiversity). The socio-economic system includes people who manage the ecological system and intervene in it (farmers) as well as the artificial means used to produce food, fiber, or other agricultural products.

The agricultural ecosystem is designed and managed by humans for the primary purpose of producing biomass. The farmer influences the nature and functioning of the ecosystem through two types of practices:

- practices that determine the configuration of the agricultural ecosystem and therefore the nature and potential volume of agricultural outputs for a given climate. This includes the choice of plant and animal genotypes (species, varieties, breeds); sowing dates and density; crop rotations; and animal presence in the ecosystem (role of grazing in the livestock feeding strategy);

- biomass production management practices:
  - limiting abiotic stresses (e.g., through water and mineral element supply) or modifying the physico-chemical conditions of the soil (e.g., through tillage or liming);
  - reducing biotic stresses (e.g., through the use of pesticides);
  - exporting the plant biomass from the field (harvesting) or return it to the soil.

The composition and functioning of agricultural ecosystems are different from "semi-natural" ecosystems because of the interaction between two components of biodiversity: so-called "planned" biodiversity and "associated" biodiversity. Planned biodiversity includes the cultivated plant species (annual, semi-annual, or perennial) and livestock intentionally introduced into the ecosystem for agricultural production purposes. Associated biodiversity includes weed species present in the field, soil fauna (endogenous macro- and meso-fauna, soil microbial communities), and the aboveground and airborne macro- and meso-fauna moving through the field and its immediate environment. The structure and dynamics of associated biodiversity depend on the planned biodiversity (plant notably serving as a food source and habitat for animal biodiversity); on biomass management practices; and on the structure of adjacent ecosystems (e.g. the composition and configuration of semi-natural habitats or forested areas).

NB: In this book, the term "agricultural ecosystem" is frequently used in the singular form to designate the ecosystem type that is the focus of EFESE-AE. Nevertheless, this should be understood to refer to "the totality of agricultural ecosystems in all their diversity."

▌ A framework for the description and assessment of ecosystem services

Many conceptual frameworks link the concepts of ecosystem structure and biophysical processes, ES and benefits along a chain (or cascade) that links ecosystem functioning to human well-being. An international literature review published in 2012 identified two major types of definitions: i) those in which ES are understood as biophysical components of ecosystems, from which benefits are derived – the definition adopted by the authors of the CICES[9]; and ii) those in which ES are understood as the benefits received by humans from ecosystems – the definition adopted in the MEA report.

Given the focus on agricultural ecosystems, EFESE-AE followed the CICES, choosing to understand ES as ecosystem "components" from which humans derive benefits that

---

9. The Common International Classification of Ecosystem Services (CICES) was introduced in 2009. CICES was developed within the context of work being done by the European Environment Agency and the United Nations Statistical Division seeking to revise and standardise the international system of environmental accounting (System of Environmental Economic Accounting – SEEA).

contribute to improving their well-being. This understanding clearly distinguishes the concepts presented in the following pages, and summarised in Figure 1-1.

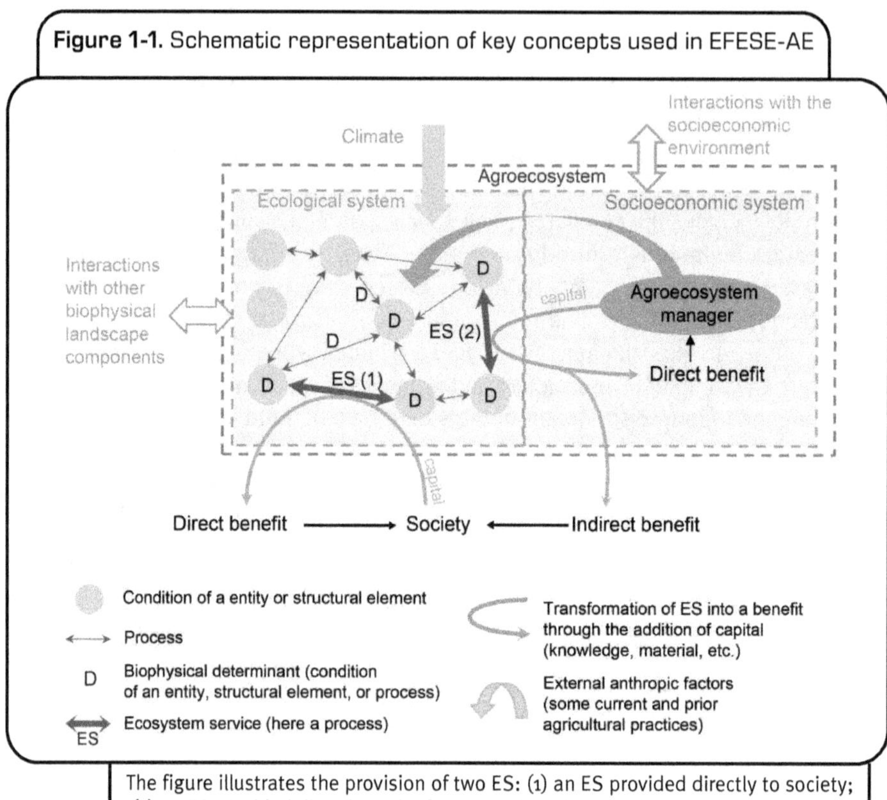

Figure 1-1. Schematic representation of key concepts used in EFESE-AE

The figure illustrates the provision of two ES: (1) an ES provided directly to society; (2) an ES provided directly to the farmer, with society receiving an indirect benefit.

Benefits and beneficiaries of ecosystem services

ES are ecological processes or ecosystem structural elements from which humans derive benefits, whether actively, by mobilising material (energy, water, crop protection products) and/or cognitive capital (knowledge, including agricultural practices), or passively (for example, benefits received from ES of climate regulation). The benefits received from ES are no longer part of the ecosystem; they may be material (goods) or immaterial (socioeconomic services[10]) in nature. A single ES may be the source of multiple benefits.

---

10. The terms *goods* and *services* are used here in the national accounting sense, and refer to the totality of goods created by businesses, public agencies, or other organisations. A *service* in the national accounting sense is different from an ecosystem service as discussed in this study.

From a public policy perspective, identifying the specific benefits obtained from ES by different groups of individuals within society can help clarify the stakes involved and highlight associated action levers for ecosystem management. Two categories of beneficiaries were distinguished here: farmers, and society as a whole. As managers of agricultural ecosystems, farmers derive from certain ES specific benefits that contribute directly to agricultural production: it is thus considered that these ES provide a direct benefit to farmers. Society as a whole is a beneficiary of ES supplied by agricultural ecosystems, both directly (in the case of ES of global climate regulation, for example), and indirectly through farmers (for example, in the case of regulating ES that substitute for the use of chemical inputs that can contaminate the environment). In the second case, the way in which society benefits from ES depends on farmers' behavior.

We should note that as residents and citizens, farmers also belong to the second category of beneficiaries, society as a whole. Given the thematic focus of EFESE-AE, other categories of social actors were not considered.

Ecosystem services, biophysical determinants and external factors

Ecosystems are made up of an assemblage of biotic and abiotic entities and processes in interaction. The structure of the ecosystem is defined by the nature and the interrelationships of those entities (spatial pattern, functional interactions...). The structure of the ecosystem and the condition of its various entities determine ecological processes (e.g., population dynamics, competition among populations) and vice versa. For example, processes of predation or parasitism determine the condition and the structure of pest species communities, which in turn determine the nature of these processes and the degree of damage to crops.

ES are the sub-group of processes or entities from which humans receive direct benefit(s). The level of ES provision thus depends on the condition of ecosystem entities and on overall ecosystem functioning. In the analytical framework of EFESE-AE, the principal ecosystem entities and biophysical processes determining the level of ES provision are referred to as "biophysical determinants." For example, the ES of pollination corresponds to the process of pollen transfer between male and female flowers. The characteristics of pollinator communities (structure, composition, abundance) are major biophysical determinants for this ES. Note that only those processes involving living organisms are recognised as ES (as a corollary, biodiversity is a biophysical determinant of ES). Certain abiotic entities or processes (e.g., soil texture) can be considered to be biophysical determinants when their interactions with biotic entities or processes determine the level of ES.

In addition, certain processes external to the ecosystem, both natural (i.e., climate) and anthropic (human activities), can increase or decrease the level of ES provision, directly and/or through their effects on biophysical determinants. These are referred to as "external factors" in EFESE-AE. For example, the ES of nutrient supply to crop plants is influenced by phenomena connected to climate change as well as by fertilisation practices; these factors directly influence levels of soil organic matter, one of the key biophysical determinants for this ES.

In the case of agricultural ecosystems, agricultural practices can play various roles. Insofar as they define the nature of the agricultural ecosystem, ecosystem configuration practices help determine the level of ES provision. Biomass management practices are considered as external anthropic factors when they influence the level of ES provision. Such practices can exert an effect either through their historic influence on ecosystem conditions (e.g., effects of tillage practices on soil organic matter) or through their effect on ES expression within the assessment timeframe (e.g., crop protection practices, which, through their effects on pest species and their natural enemies, influence the level of ES for pest species regulation over the course of a year).

Ecosystem services, "dis-services" and impacts of agricultural practices

In the literature, the concept of "dis-services" is often used to refer to two distinct phenomena: 1) the negative effects of some biodiversity components or certain ecosystem processes on human well-being; 2) the negative impacts of human activities on the environment.

First of all, one needs to distinguish between the negative effects of some types of ecosystem functioning on human beings and situations of low ES provision. Negative effects of ecosystem functioning (dis-services of type 1) can include some effects of wild fauna in agroecosystems or in urban areas; or the release of pollen allergens by plants. A low effective level of ES corresponds to a situation in which a low level of benefit is received.

Furthermore, transposed to the case of agricultural ecosystems, dis-services of type 2 correspond essentially to material flows from agricultural ecosystems to other ecosystems as a result of agricultural practices. Thus, some biomass management practices (e.g., crop protection treatments, fertiliser applications) create pollution (e.g., pesticides, nitrates) that moves beyond the agricultural ecosystem and ultimately results in a reduction of human well-being. ES and the negative environmental impacts of human activities offer two different and complementary ways of looking at ecosystem functioning. For example, the conversion of N_2O (a greenhouse gas) into N_2 is an ES, whereas N_2O emissions resulting from nitrogenous fertiliser applications are an environmental impact.

When assessing ES and dis-services, it is important to keep in mind that a single ecological process may be considered an ES for one category of beneficiaries and a dis-service for another: the definition of ES and dis-services thus depends on the group of actors considered. For example, the regulation of wild ungulates by large predators may be an ES for foresters, but a dis-service for hunters or hikers. Dis-services were not examined in the study that gave rise to this book. Nevertheless, where useful and appropriate, indicators of the negative impacts of agricultural practices were developed and quantified in addition to levels of ES provision (e.g., fixed N vs. leached N; see Chapter 6).

Identifying and assessing ecosystem services from agricultural ecosystems

▌ Typology of ES provided by agricultural ecosystems

In keeping with much of the international research and the EFESE program, EFESE-AE used the CICES classification (version 4.3[11]) as its reference typology for the identification of ES from agricultural ecosystems. CICES classifies ES into three major groups:
- "provisionning services," corresponding to the production of biomass, water, and energy by the ecosystem;
- "regulation and maintenance services," corresponding to ecological processes that help regulate phenenoma such as the climate, the frequency and magnitude of disease outbreaks, or various aspects of the water cycle (e.g., floods, water quality) and the movement of material by water (e.g., erosion);
- "cultural services," source of recreational, aesthetic, or spiritual benefits.

After selecting for analysis a group of ES supplied by agricultural ecosystems in France among the ES listed by CICES,[12] the expert group determined for each one, based on the international scientific literature: i) the nature of the ES, ii) the benefits received by society and (where applicable) by farmers; iii) the major biophysical determinants and external factors involved in ES provision. This work led the group to refine and adjust the classification of some of these ES, and thus to significantly revise the CICES typology (cf. Annex 2).

In particular, the status of agricultural production is widely debated in the scientific literature the international ES assessments. Agricultural production results from interactions between regulating ES and anthropic inputs (energy, irrigation, fertilisation, pesticides). Treating agricultural production as a "provisioning service" suggests that a higher level of this ES can result from an increased use of external inputs, not just from better ecosystem functioning. In order to distinguish between the respective roles of ecosystem functioning and external input supply in agricultural production, the concept of a "provisioning service" was not adopted in EFESE-AE. Instead, agricultural production was understood as an agricultural good, or in other words, as a benefit received by the farmer from interactions between certains regulating ES – called "input services" in EFESE-AE – and anthropic external inputs (see Chapter 3). In addition, the "water supply" ES as defined by the CICES were understood as water flow regulation ES (see Chapter 2).

The definition of so-called "cultural" services is likewise the focus of some debate. In practice, the majority of items identified in this category by the CICES correspond more closely to a typology of landscape uses and/or values (and thus benefits) than

---

11. The most recent version when this study was completed - https://cices.eu/resources/
12. ES marginal within the French context or supplied exclusively by other types of ecosystems (for example, regulation of salt water quality) were excluded from the scope of the study. In addition, some ES supplied by agricultural ecosystems could not be examined due to a lack of relevant expertise among the study team.

to ES in the sense adopted in EFESE-AE. As a result, only those services described as "recreational" in the CICES framework were examined here (cf. Chapter 4).

Table 1-1 presents the final list of ES examined within EFESE-AE. All the ES were the subject of a review of the scientific literature to propose an evaluation methodology. Where possible, the expert group used these methods to obtain a biophysical or even economic quantification of the ES. Where this was not possible, they identified the need for additional work and data that would enable them to be implemented.

Table 1-1. Final list of ES examined in EFESE-AE

ES Designation	Biophysical analysis	Economic analysis
Pollination of crop plants	Quantified	Assessed
Regulation of weed seeds	Partially quantified	Investigated
Regulation of insect pests	Partially quantified	Investigated
Soil stabilisation and erosion control	Quantified	Investigated
Soil structuration	Investigated	/
Storage and return of water to crop plants	Quantified	Assessed
Storage and return of blue water	Quantified	/
Supply of mineral N to crop plants	Quantified	Assessed
Supply of other nutrients to crop plants	Investigated	/
Natural attenuation of pesticides by soils	Investigated	/
Regulation of water quality with respect to N, P, and DOC	Partially quantified	Investigated
Global climate regulation by GHG attenuation and C sequestration	Quantified	Investigated
Recreational potential (outdoor activities, no sampling)	Partially quantified	/
Recreational potential (outdoor activities, with sampling)	Investigated	/

Some other ES strongly relevant for to agricultural ecosystems should be studied to supplement the work carried out in EFESE-AE. This is particularly the case for:
- regulation of crop and livestock diseases;
- decomposition and the recycling of dead matter and waste products;[13]
- local climate regulation (at the landscape or field level);
- regulation of air quality;
- regulation of flood.

---

13. Performed by necrophagous and coprophagous organisms, respectively; primarily insects (but also carrion-eating birds in the case of larger carcasses).

▌ Biophysical assessment of ecosystem services

Scope of EFESE-AE

Agricultural ecosystems are defined here as land areas used primarily for agriculture. The biophysical compartment under examination is the soil-plant-animal system. The agricultural ecosystem is mainly made up of cultivated or grassland fields, considered as the functional units of the ecological system, and semi-natural elements located within and around the fields (field borders, roadside margins or ditches, groves of trees, ponds, hedgerows, grass strips). The agricultural ecosystem exists within a landscape mosaic of interacting ecosystems; that is, ecosystems that exchange matter and energy, notably as a result of the movements of associated biodiversity. In areas where the most heavily represented ecosystem within this mosaic is agricultural in nature, we speak of an "agricultural landscape."

Horizontal delimitation

French agricultural area – including arable land plus land in permanent grass or perennial crops – accounts for approximately 29 Mha, or 54% of the total land area of France. Figure 1-2 shows agricultural area as a percentage of total land area for each Small Agricultural Region (SAR).[14]

SAR in which agricultural area accounting for less than 25% of total land area are located in the Landes of Gascony (primarily forested); the Sologne, in north-central France (dominated by wooded and wetland areas); the combined area formed by the Vivarais, the Cévennes and the Causse du Larzac; and mountainous areas (the Vosges, the Alps, the Pyrenees, and Corsica).

Figure 1-3 shows the percentage of woody formations located within the agricultural fields for each SAR in 2012 as reported by farmers in the French LPIS.

The highest percentages are observed in mountainous areas (the Alpes, the Pyrenees, the Jura, the Vosges, the Massif central), in the armorican Massif (Brittany, Lower Normandy, the Loire Valley), and around the Mediterranean (Languedoc).

---

14. French statistical zoning that divides the national territory into homogeneous areas in terms of soil type, climatic conditions and dominant agricultural production.

Figure 1-2. French agricultural area as a percentage of total land area in each each SAR in 2010

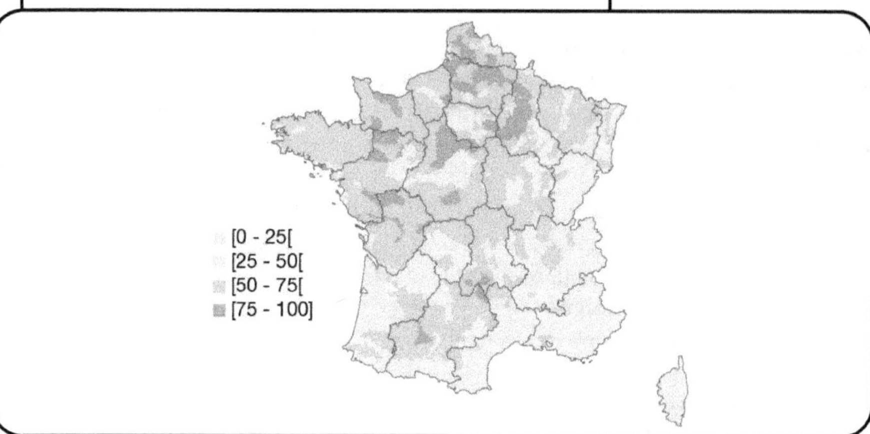

Sources: French agricultural census (RA, 2010) and data from National Institute of Geographic and Forest Information (IGN, 2010).

Figure 1-3. Percentage of woody formations in the agricultural fields of each SAR in 2012

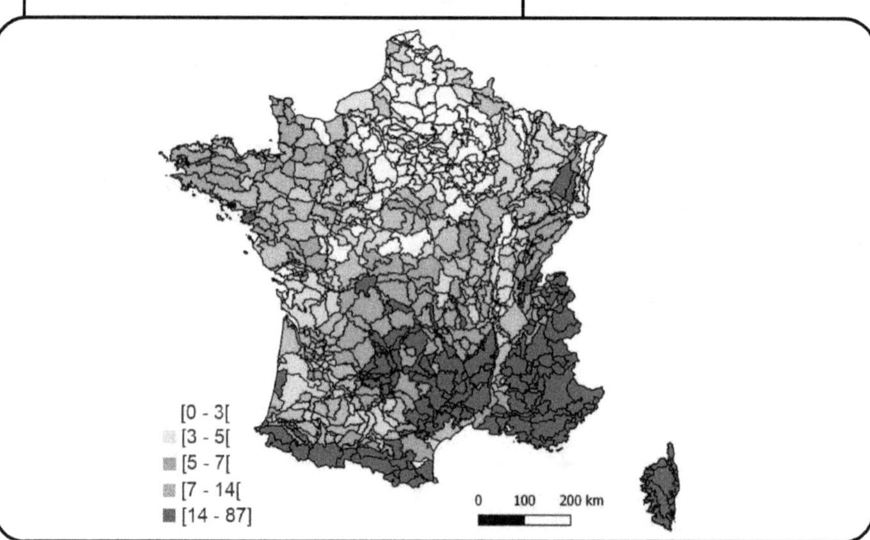

Sources: data from the French Land Parcel Identification System (RPG, 2012) and the vegetation layer of the National Institute of Geographic and Forest Information topographic database (IGN BD TOPO®).

Vertical delimitation

From the perspective of ES assessment, the vertical reach of the agricultural ecosystem (Figure 1-4) was considered to extend from the top of the saturated soil zone to the top of the plant canopy (the saturated soil layer is addressed by geo-hydrological research not examined for this study).

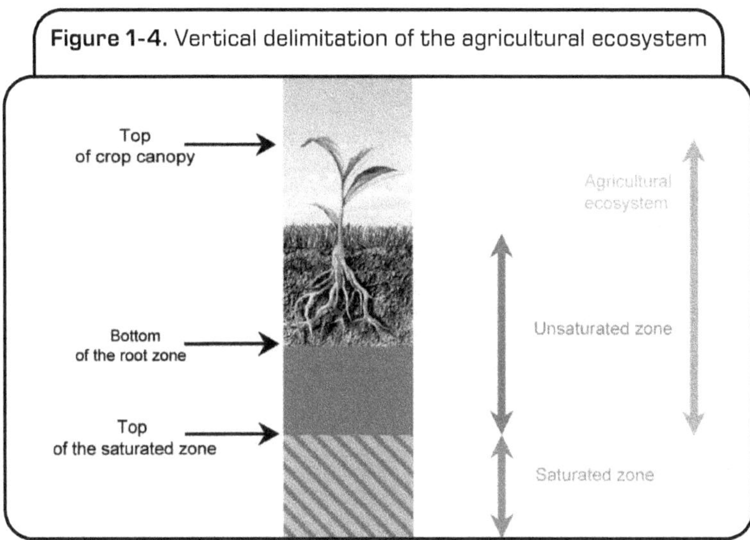

Figure 1-4. Vertical delimitation of the agricultural ecosystem

General characteristics of biophysical assessment

Based on an analysis of the scientific literature and European assessments on ES (notably as led by the European Commission's Joint Research Center and the MAES working group), the expert group identified available indicators and data for quantifying the level of provision of each ES. These methods should make it possible to map ES from agricultural ecosystems i) as they are currently managed (in order to have an overview of the levels of SE currently provided), ii) at the highest spatial resolution; and iii) accross the entire territory of mainland France. Depending on the nature of the processes analysed, the availability of data (Box 1-1) and the technical limitations of the quantification tools, each (group of) ES was therefore assessed using its own method and spatial resolution.

Two main types of approach were used.

- ES supplied by the functioning of the soil-plant-animal system in the field (e.g., supply of mineral N to crop plants) were quantified using dynamic models that simulate fluxes of C, N, water and energy in the ecosystem. Two models were used, both developed by Inra: STICS for major field crop systems, and PaSim for grassland systems (see Chapter 2, Box 2-1). These ES were quantified on the scale of spatial units homogeneous in terms of soil and climate characteristics (the French agricultural area was divided into 24,356 units called PCU).

• ES determined by landscape characteristics (e.g., pollination of crop plants) were quantified by using indicators of landscape composition and configuration. Depending on the method, these ES were quantified for each 100 m square cell, 2 km square cell or for each Department.[15]

Box 1-1. Main databases used to quantify ES levels

Representation of the spatial distribution of land use categories was based primarily on two databases:

• The French LPIS[16] database was used to characterise the nature and geographic extent of agricultural ecosystems. Since this database includes only those agricultural land areas eligible for Common Agricultural Policy payments, some types of ecosystems are poorly represented, notably vineyards and orchards. As a result, the assessments apply primarily to major field crop and grassland areas (prairie, summer pastures), which together currently account for 95% of French agricultural area;

• The vegetation layer of the National Institute of Geographic and Forest Information (IGN) BDTOPO® database was used to characterise semi-natural elements situated within field areas (hedges, isolated trees, etc.).

A database developed by Inra from an analysis of the French LPIS for 2006 through 2012 (Leenhardt *et al.*, 2012)[17] provides crop rotations for each crop block in mainland France.[18] This database was used whenever the evaluation of SE required an analysis of the sequence of crops or grasslands.

Data on farmers' management decisions (e.g., length of grazing, animal stocking rates) and agricultural practices were drawn for the most part from two surveys conducted by the French Ministry for Agriculture:

• the Agricultural Census 2010, providing data at the level of the canton[19] or at the level of the SAR.

• the Agricultural Practices Surveys from 2006 and 2011, providing data at the level of the administrative region (for the year of the survey). This database was used for information on agricultural practices with respect to seeding and fertilisation.

Finally, most of the maps were generated excluding non-agricultural ecosystems and urban areas. Since most assessments of ES and goods are focused on major field crop and grassland ecosystems, most of the maps also exclude other agricultural ecosystems. Masking those areas was achieved by crossing the French LPIS data with CORINE Land Cover (CLC) data.

---

15. NUT 3 level in the European Nomenclature of Territorial Units for Statistics.

16. The Land Parcel Identification System (LPIS) was designed as the main instrument for the implementation of the Common Agricultural Policy first pillar, whereby direct payments are made to the farmer once the land and area eligible for payments have been identified and quantified. As defined in Art. 70 of Regulation (EU) No 1306/2013 and Art. 5 of Regulation (EU) No 640/2014), each Member State has to establish his proper LPIS tool. In France, this is the RPG for *Registre Parcellaire Graphique*.

17. Leenhardt, D., Therond, O., Mignolet, C., 2012. Quelle représentation des systèmes de culture pour la gestion de l'eau sur un grand territoire? *Agronomie, Environnement & Sociétés*, 2 (6), 77-90.

18. A crop block is a grouping of contiguous fields homogeneous in terms of cultivation, cropping history (succession of crops and fertiliser inputs) and the nature of the soil.

19. Territorial subdivisions of the French Departments.

All types of French agricultural ecosystems were *a priori* of interest. However, apart from a few exceptions, ecosystems cultivated with perennial crop (vineyards, orchards, perennial energy crops) or dedicated to market gardening, and those located in the French overseas territories (banana plantations, etc.) were not examined due to a lack of data necessary for their characterisation.

Finally, the effect of interannual variations in climate and in ecosystem configuration (crop rotations) on average levels of ES provision was accounted for by using temporal data series wherever possible. Most of the results presented in the following chapters thus correspond to annual averages.

Appendix III summarises the main characteristics of the evaluation carried out for each quantified SE (nature of the indicator, basis of the assessment, etc.).

Beyond the biophysical quantification of ES, identification of the ecosystem contribution to agricultural production represents an active area of research. A number of original methodological proposals were developed and implemented in EFESE-AE. Preliminary results for the production of plant goods (crops, forages) and animal goods (meat, milk, eggs, livestock) are presented in Chapter 3.

▌ Economic assessment of ecosystem services

The idea of "ecosystem services" represents one link in a chain of concepts connecting the condition and the functioning of natural systems to human well-being. Whereas the biophysical analysis of ES seeks to understand interactions between ES and ecosystem functioning, the economic approach addresses the corresponding link between ES and human well-being. While the two approaches share the central concept of benefit, connecting the biophysical and economic assessments is recognised as a key challenge in the scientific literature. In practice, the biophysical approach usually produces ES indicators that are not directly or easily useable by economists. The economic approach, furthermore, often seeks to quantify benefits derived from ES, rather than ES themselves, as a way of assigning a value to the ES. As explained in the beginig of this Chapter, the assessment of ES, functionally connected to the ecosystem, corresponds poorly to the assessment of benefits, functionally disconnected from the ecosystem and relating instead to the socio-economic subsystem. The assessment of benefits requires accounting for the capital inputs (material, human, institutional, financial) used to obtain these benefits from ES.

Identifying the benefits received by human beings from ES is also not an easy matter. "Society" as an ES beneficiary is in reality composed of a multitude of economic agents who do not all benefit equally or in the same way from the ES, notably as a result of economic interactions among these agents. For example, the farmer benefits directly from the ES "supply of mineral N to crop plants;" the benefit he or she receives is a savings in the use of synthetic N fertilisers. Indirectly, at the level of society as a whole, the farmer's enjoyment of the ES allows him or her to avoid the environmental impacts associated with the use of synthetic mineral N. As a result, it was decided to limit the analysis to the benefits received directly from ES by society as a whole or by the farmer.

The work consisted of 1) the description of a list of agricultural goods and ecosystem services; 2) their biophysical assessment (quantification of the level of provision); and, when possible 3) the economic assessment of ES. The economic assessment is little developed in comparison with the biophysical assessment. It highlights the issues and difficulties specific to this assessment and the research prospects that need to be developed in order to complete this work.

2. Input ecosystem services: regulating services that support agricultural production

The primary goal of agricultural ecosystems is to produce plant and animal goods. Plant production results from interactions between regulating services and external inputs supplied by the farmer (energy, irrigation, fertilisation, pesticides). As a result, agricultural goods may be considered as "co-produced" by the ecosystem and human activities.

Regulating services that determine plant production can be considered as factors of production. Some studies accordingly refer to them as "input" services. As the manager of the agricultural ecosystem, the farmer is the direct beneficiary of these ES. In addition, when they replace synthetic inputs, input services contribute indirectly to the reduction of environmental contamination, creating a benefit for society as a whole.

 ## Input ecosystem services and agricultural production

The agricultural ecosystem is modified and managed by the farmer with a view to (i) introducing the plant and animal species whose biomass will be exported from the agricultural ecosystem following a period of growth, and (ii) altering soil physicochemical characteristics and other abiotic and biotic stresses that would otherwise impede expected production levels (lack of water, lack of nutrients, insufficient pollination, pest pressure, etc.). In addition to external inputs supplied by the farmer (soil tillage, synthetic inputs, etc.), a certain number of regulating services will influence crop yield formation by affecting various factors that can limit or reduce it (Table 2-1). These ES include:

- ES involved in soil fertility, determining soil physicochemical characteristics and thus abiotic stress levels;
- ES involved in "biological" regulation – that is, ES relating to aboveground associated biodiversity, including insect pollination (involved in the sexual reproduction of approximately 2/3 of cultivated species) and the regulation of crop pests.

In interaction with ecosystem characteristics, regulating ES, and agricultural practices, climate is involved in determining the potential agricultural productivity of a given location. Indeed, the level of expression of the whole suite of processes at work within the agricultural ecosystem – input services among them – is linked to climate.

Table 2-1. Agricultural practices and input services contributing to the control of abiotic and biotic ecosystem characteristics that limit or reduce crop yield

	Agricultural practices	Regulating input services
Abiotic characteristics:		
	→ Soil tillage	→ Soil structuration
	→ Inputs of organic and mineral fertiliser	→ Supply of nutrients to crop plants
	→ Irrigation	→ Storage and return of water to crop plants
	→ Practices to improve and protect soil integrity	→ Soil stabilisation and erosion control
	→ Other practices with multiple objectives: return of biomass (crop residues) to the soil, inputs of organic amendments	
Biotic stresses:		
Damage caused by weeds and crop pests	→ Inputs of crop protection products → Biological control through the introduction of organisms not naturally present in the ecosystem or input of plant health stimulants → Mechanical weed control	→ Conservation biological control: regulation of weed seeds, regulation of pests
Pollination deficits	(→ Manual pollination)	→ Pollination of crop plants

Note that at the level of the field unit, climate can also be influenced by the structure of the agricultural ecosystem and its surrounding landscape matrix. By locally modifying certain climate parameters (temperature, air displacement, etc.), these structural elements can create a microclimate that is more or less favorable to certain processes, and thus to the expression of some regulating ES. This local climate regulation can be understood as an ES provided by the agricultural ecosystem.

Table 2-2 presents the list of input services considered in EFESE-AE and the nature of the analysis conducted for each. Where possible, the level of ES supply was quantified. Two input services known to contribute to agricultural production, the regulation of crop diseases and the regulation of local climate, were not examined because they were determined to require specific disciplinary competencies not represented on the expert team. The table also presents the main benefits that beneficiaries can derive from these ES. By definition, the farmer benefits directly from these ES; society benefits essentially indirectly.

Table 2-2. Input services examined in EFESE-AE

Input ecosystem service	Direct benefits received by the farmer	Direct and *indirect* *(in italics) benefits* received by society	Biophysical analysis
Soil structuration	Reduction in soil tillage		Methodological avenues
Storage and return of water to crop plants	Reduction in the amount of water supplied by irrigation	*Reduction in water deficits linked to irrigation*	Quantified
Supply of mineral N to crop plants	Reduction in the amount of nutrients supplied in the form of external fertilisers	*Reduction in pollution linked to the use of fertilisers*	Quantified
Supply of other nutrients to crop plants			Methodological avenues
Soil stabilisation and erosion control	Preservation of the agronomic potential of the field	Reduction in mudslide phenomena; Improvement in surface water quality *via* the reduction of waterborne soil particles	Quantified
Pollination of crop species	Increase in crop production and yield stability over time; Avoidance of costs for alternative pollination strategies (hive rental, manual pollination).		Quantified
Conservation biological control: (regulation of weed seeds and crop pests	Protection of yields; Reduction in the use of crop protection products and/ or mechanical weed control operations	*Reduction in pollution linked to pesticide use*	Quantified

 Regulation of soil physicochemical characteristics

▌ Soil structuration

The CICES classification proposes two sub-groups of regulating ES linked to the formation and maintenance of soil biogeochemical conditions. The first includes processes relating to the maintenance of soil fertility, nutrient storage, and the maintenance of soil structure, and including processes of soil formation and alteration (biological, chemical, and physical). The second relates to the decomposition / mineralisation of organic matter, nitrification / denitrification, N fixation, etc. This typology is not well suited to the analytical framework presented in Chapter 1, however. Given the diversity of processes included in each category, it is difficult to attribute human beneficiaries and received benefits to the ES thus defined. Moreover, some of the processes appearing in the CICES typology, such as pedogenesis, take place on time scales much longer than that of the year (or the crop rotation). This makes it difficult both to assess an ES defined by such processes and to include it in a multi-service analysis approach. Finally, the processes involved in the regulation of the cycle of mineral elements do not appear clearly within this typology, whereas they are among the most frequently cited processes in the scientific literature on soils.

The majority of academic work that deal with ES relative to soils mention a specific ES relating to soil structure or soil formation. Soil structure, defined as the organisation of its liquid, solid, and gaseous components, is a key property of soils. Soil structure is identified by many authors as an essential determinant of soil quality and soil health, and as a fundamental ecological indicator of soil condition. The earliest publications on ES in connection with soils thus define an ES for "soil structure," although this is more suggestive of soil condition at a given moment than an ES in the strict sense.

For EFESE-AE, therefore, the CICES typology has been redefined to distinguish soil structuration from ES relating to nutrient cycling. In keeping with several recent publications, "soil structuration" ES is defined here as the ecosystem's capacity to generate and maintain a soil structure allowing the soil to fulfill its functions of support, habitat, filter, and storage. From an agricultural perspective, this ES directly benefits the farmer because it reduces mechanical operations for soil structuration (e.g., plowing).

Biophysical determinants and external factors

Biophysical determinants

The structural condition of the soil results from physical and biological processes that lead either to the creation of air spaces within the soil (fissuration, perforation, etc.) or to their disappearance (rain impact, compaction, etc.).

There is a strong connection between soil structure and soil biological activity. Soil structure determines key characteristics of faunal (macrofauna, mesofauna, microfauna) and floral habitat, and these in turn modify soil structure. So-called "soil engineers" (earthworms, ants, termites, etc.) play a key role in soil perforation and soil aggregation. Plant covers likewise have both a mechanical effect (through root penetration) and a biochemical effect (through micro-aggregation by root exudates) on the various soil horizons. At the microscopic level, microorganisms also contribute to process of soil aggregation.

Among abiotic determinants, soil texture and soil organic matter (OM) content are key properties determining soil structural stability and soil porosity.

External factors

Soil structure is also continually subject to external factors that can create or eliminate air spaces within soils. The alternate wetting and drying of soils through climate effects can lead to soil cracking over the course of a season or over the course of a year. On the more immediate term, a soil crust can develop in several hours under the effects of a heavy rain.

Agricultural practices play an essential role in soil structural evolution, often with instantaneous effects, and sometimes also with secondary, long-term effects. In addition to the specific mechanical operations performed in the field, the passage of heavy equipment can lead to soil compaction at the surface or at deeper levels, which can persist over the medium or long term if it is not counteracted through the use of soil tillage practices to restore soil porosity. Finally, other practices can affect processes of soil aggregation and disaggregation, such as the maintenance of crop residues on the soil surface and the addition of organic fertilisers. It is also useful to consider soil structure from a multi-year perspective, since agricultural practices can negatively impact ES level over the short term but then have a positive effect over the long term, once biological and climatic processes to rebuild soil structure take hold.

Methodological avenues for quantifying levels of ES supply

Since the MAES program follows the CICES typology for ecosystem services, it does not specifically examine an ES for soil structuration. Several authors have proposed assessing ecosystem condition resulting from the ES rather than the level of ES supply itself. These approaches make use of indirect indicators of soil structure (e.g., OM content, earthworm abundance, or the presence of microarthopods, etc.). These indicators were not used in EFESE-AE because they do not allow the level of SE to be estimated.

ES assessment methods using mechanistic (process-based) modeling approaches exist, but they have never been applied at a large scale. For example, the MOSES model (MOdular Soil Erosion System), which is used to evaluate various ES, supplies an indicator of the change in soil bulk density over time. Other authors use the SPASMO model (Soil-Plant-Atmosphere) to generate ES indicators relating to soil structure. The determinants of the simulated changes are not specified in either of these two cases, however. A more advanced proposal consists of using the CAST model (Carbon Dynamics and Soil Stability) to predict the temporal dynamics of parameters that depend on soil structure, such as gas diffusivity or water permeability.

A final type of approach consists of developing statistical relationships among different soil characteristics linked to soil structure. These relationships (pedotransfer function) can make it possible to compare the effects of different cropping systems or agricultural practices within a given agro-pedoclimatic context, although they cannot be extrapolated across all situations. Other authors have proposed characterising soil structure dynamics based on the development of compacted areas within the cultivated soil horizon. Measuring this indicator over a large area would require a heavy investment in data collection in the field, however.

▋ Supply of nutrient elements to crop plants

In agricultural ecosystems, some of the mineral elements taken up from the soil by plants will be exported from the field with the harvest. Farmers supply mineral and organic fertilisers to compensate for these exports. In addition to N, the mineral elements most frequently supplied as synthetic fertilisers are P, potassium (K), sulfur (S), and magnesium (Mg). Among these elements, the most significant agricultural and environmental issues are associated with N and P flows. Although current strategies for N fertilisation, along with advances in plant breeding, have made it possible to improve the efficiency of plant N use and to limit losses to the environment, excess N linked to agricultural activities remains high. Both N and P contribute to the eutrophication of aquatic environments. In addition, the rocks from which phospate fertilisers are manufactured come from mines found in just a few countries, and represent a finite and non-renewable resource. The availability of P for global food production is thus a major issue over the medium to long term. Limiting the use of mineral N and P fertilisers – and assessing the capacity of agricultural ecosystems to supply those nutrients instead – thus represents an important global agricultural challenge.

Issues related to N are widely considered within the ecosystems services literature, but P is less often considered from this perspective. The review of the scientific literature relating to ES for nutrient supply to crop plants found three major types of approaches: i) overall considerations of the regulation of the nutrient cycle or the N cycle; ii) analyses of the processes by which N or P are made available for plant growth; and iii) assessments of the retention of N and P (viewed as pollutants) by the ecosystem. The first approach is too broad, potentially encompassing several ES, which does not make it possible to precisely identify the benefits resulting from such services. The second is the only approach that can be connected to the capacity of a soil within an agricultural ecosystem to supply N or P to crop plants. The third approach, which is the one employed by the MAES program, corresponds to the regulation of nutrient losses into waterways, an ES which directly benefits society as a whole (see Chapter 4).

Given the specific issues associated with N and P in agriculture, two ES were described in EFESE-AE: i) the supply of mineral N to crop plants; and ii) the supply of other nutrients (notably P) to crop plants. These ES enable the farmer to reduce inputs of exogenous fertilisers while maintaining the same levels of production.

Biophysical determinants and external factors

Biophysical determinants

The principal biophysical determinants of the ES for the supply of N and P to crop plants are ecosystem properties and processes affecting the quantity, chemical form (bioavailability), and physical accessibility to plants of these elements:

- total N and total P in the soil;
- abiotic and biotic soil properties and processes that determine the amount of N and P present in different forms (mineralisation and organisation by soil microbiota, affinity of the soil's solid phase to phosphate ions);
- soil porosity and soil moisture levels, which determine nutrient diffusion to plant roots in bioavailable form (mineral N and dissolved P) in the soil's liquid phase;
- processes of N and P assimilation by the crop.

External factors

External factors that can affect the ecosystem's capacity to supply these ES are those that relate to soil physicochemical properties and soil biological activity. Climate affects soil temperature and soil moisture levels, which in turn have a direct impact on soil biological activity (notably the mineralisation of OM). Fertilisation practices (mineral and/or organic) and methods of crop residue management determine N and P inputs into the soil and the dynamics of soil processes noted above. Finally, irrigation and soil tillage practices impact the level of ES because they have effects on moisture levels, temperature, soil structure, etc.

Level of ES provision: the case of N

Given the available data and tools (particularly modeling tools), only ES for the supply of mineral N to crop plants could be quantified in EFESE-AE. Methodological avenues for assessing the ES for the supply of P are presented in the full report of the EFESE-AE study.

Numerous authors have proposed using an indicator for the quantity of N present in the soil as a way to assess the level of ES supply. However, this variable is not an adequate proxy for the ES since both mineral N availability and crop N requirements vary significantly over the course of the year. More direct indicators of the potential level of ES provided by the agricultural ecosystem are the quantity of mineralisable N or the potential rate of N mineralisation. These can be measured in the laboratory but they are not currently the subject of a quantification protocol on all French soils.

Assessment method

The evaluation is based on the use of two models of plant cover dynamics (STICS for major field crop ecosystems, PaSim for grassland ecosystems – see Box 2-1) that simulate the different components of the N balance:

$$N_{soil} \text{ at harvest} - N \text{ removed} = N_{soil} \text{ at planting} + N \text{ supplied}$$

N removed corresponds to the amount of N exported in the form of harvested biomass and losses to leaching and volatilisation. N supplied corresponds to N fixed by symbiosis, the mineralisation of soil organic matter (OM) and crop residues, and inputs of external N fertilisers.

Calculating the quantity of N exported in the form of harvested crops makes it possible to quantify the level of ES effectively supplied by the agricultural ecosystem, provided that the N used by the crops comes exclusively from mineral N supplied by the agricultural ecosystem thanks to the net mineralisation of soil OM and residues. To make such a calculation, a simulation premise representing cropping systems without N fertilisers, all other things being equal, was tested. The results of these simulations were judged to be unuseable, however, due to a significant loss of OM over the 30 years of the simulation and thus changes in N availability resulting from OM mineralisation. Only those results corresponding to simulations of practices currently in general use (i.e., using supplied N fertiliser inputs) were analysed.

Two indicators were calculated:

• The first represents the level of ES potentially supplied by the ecosystem (as opposed to the level of ES effectively used by the farmer), and corresponds to the quantity of total mineral N supplied by the ecosystem during the period of crop growth (by symbiotic fixation and mineralisation). This indicator reflects the ecosystem's capacity to supply mineral N to the crop during its growth cycle, taking into account the initial state of the ecosystem (at the time of sowing) and its interannual evolution (due to cumulative effects). This indicator is an imperfect proxy since the supply of external fertiliser inputs is indirectly accounted through the mineralisation of the preceding year's crop residues during the crop year.

• The second corresponds to the total quantity of mineral N available to the commercial crop *excluding* that supplied by fertilisation during the crop year. This indicator corresponds to the quantity of mineralised and symbiotically fixed N during the period of crop growth to which is added the quantity of mineral N in the soil at the time of sowing. The latter results from the application of mineral fertilisers in the preceding year, and includes any residual mineral N in solution in the soil and N from the mineralisation of crop residues in between harvest and sowing. This indicator is thus potentially more influenced by the effect of fertilisation in the previous year.

> ## Box 2-1. The use of dynamic simulation models of the soil-plant-(animal) system to assess ES relating to the C, N, and water cycles
>
> Dynamic simulation models of the soil-plant-(animal) system were used in EFESE-AE to quantify ES relating to the water, N, and C cycles: STICS[20] and PaSim[21]. These models simulate the functioning of cropping systems (soil-field crops) and permanent grassland systems (soil-grasslands-grazing animals), respectively. Major processes simulated by these models include the growth and development of plant cover, and various components of the water, N, and C balances. These two models have been the focus of previous assessments, and the expert group was already proficient in their use. A simulation plan was developed specifically for EFESE-AE. Due to the time available and other constraints, however, only the results from the STICS model could be made use of. For this reason, the ES to be assessed with this tool were only quantified for major field crop systems.
>
> The use of dynamic simulations seeks to estimate the annual average level of ES supplied by a temporal configuration of crop cover corresponding to the most widespread systems currently in use in France. In other words, the functional unit of the assessment is not the annual plant cover but the crop rotation or sequence of grassland covers. This makes it possible to account for multi-year effects (both the "prior year" effect and cumulative effects) on the average level of the ES under consideration.
>
> To obtain a suitable sample of the effects of climatic variation on the variables to be analysed, simulations were carried out over a period of 30 years (1984-2013) defined according to the availability of climate data. The goal was not to analyse past behavior, nor to examine changes in phenomena over time, nor to predict the functioning of future simulated cropping or grassland systems, but rather to assess the average performance of these systems over a sufficiently long climatic series so as to eliminate the "year" effect.
>
> **Model input parameters**
>
> The simulations were conducted over units of land area considered to be homogeneous in terms of soil and climate, called Pedo-Climatic Units (PCU). In total, 23,149 PCU containing a minimum of 100 ha of declared agricultural land in the French LPIS were considered (Figure 2-1).
>
> Input parameters were defined for the PCU using various databases (Figure 2-2):
>
> • climate was characterised using Météo France's SAFRAN database (8 km square cells);
>
> • soil properties and characteristics were supplied by the Inra US Infosol for each Soil Mapping Unit (SMU) of the soil map at the scale of 1/1,000,000 (Base de Données Géographique des Sols de France - BDGSF). A single SMU may be associated with several soil types.
>
> • the soil condition with respect to N and organic C at the beginning of the simulations were drawn from data in the literature;
>
> • crop rotations and grassland types were drawn from an analysis of the French LPIS from 2006 to 2012. A maximum of two rotations and two grassland types were selected per PCU, with the chosen systems corresponding to the system types covering the largest amount of land area within that PCU ("dominant" systems).

---

20. For *Simulateur multidisciplinaire pour les cultures standard* - https://stics.inrae.fr/eng/
21. For Pasture Simulation Model - https://hal.inrae.fr/hal-02808903

• agricultural practices were generally characterised based on data from the Agricultural Practices surveys of 2006 and 2010 and the Grasslands survey of 1998, conducted by the French Ministry for Agriculture. Due to a lack of other relevant data, irrigation practices (relevant here only for maize production) were automatically simulated by the STICS model so as to supply 85% of crop water requirements.

Figure 2-1. Distribution of PCU considered in the assessment according to the type of simulated agricultural land use

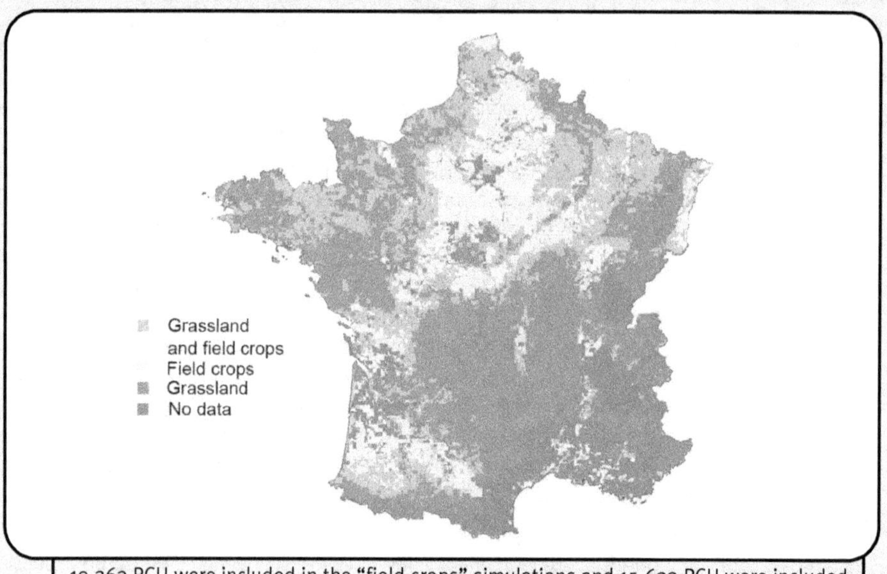

10,263 PCU were included in the "field crops" simulations and 15,623 PCU were included in the "grassland" simulations; PCU shown in gray (including Corsica): no simulation.

Figure 2-2. Simplified overview of the simulation model for crop and grassland systems in France

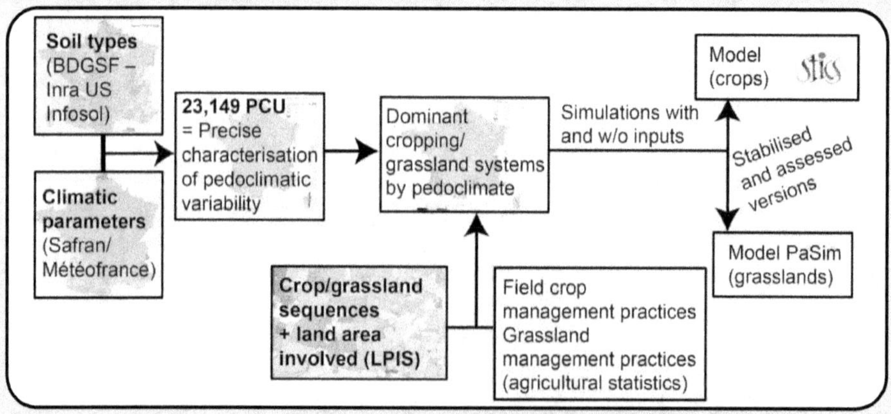

Simulation scenarios

The simulation tool thus elaborated made it possible to simulate "current systems," in other words, crop and grassland systems managed according to prevailing practices for eight crops and three types of grasslands. Practices included:

• mineral and organic N fertilisation: one or two dominant fertilisation practices were simulated for each PCU (mineral fertilisation only and/or mineral + organic fertilisation);

• mode of biomass removal in accordance with N fertilisation practices: grain removal in "field crop" systems in combination with mineral fertilisation only; removal of cereal straw and maize harvested as silage in "livestock" systems, in combination with organic fertilisation;

• incorporation of crop residues;

• irrigation of maize in PCU where the majority of land in maize is irrigated (information from the LPIS);

• inclusion of cover crops in crop rotations in PCU located in Nitrate Vulnerable Zones (in accordance with the Nitrates Directive);

• methods of grassland management: type and frequency of mowing, silage harvesting, grazing;

• livestock density: number of Large Animal Units grazing per hectare at time t based on data from the Agricultural Census 2010.

Several alternative simulation scenarios were also created in order to the test the effects of specific practices on the level of ES supply (other practices unchanged relative to the "current systems" simulations):

• alternative simulations, "maize without irrigation" in PCU that are typically irrigated;

• alternative simulations, "no nitrogenous fertilisation" for all PCU;

• alternative simulations, "no cover crops" for PCU located in Nitrate Vulnerable Zones.

Example: Consider one PCU characterised by two soil types, one covering 25% of the land area in the PCU and the other covering the remaining 75%. Two prevailing crop rotations were assigned to this PCU:

- 15% of the PCU has a rotation of spring wheat / spring wheat / sugar beet, managed with two methods of nitrogenous fertilisation: 40% of land area is fertilised with mineral fertilisers only, and 60% with organic fertilisers supplemented by mineral fertilisers;

- 85% of the PCU is in continuous forage maize, not irrigated, with mineral fertilisation only.

This PCU is located in a Nitrate Vulnerable Zone, so each of these three cropping systems is simulated with and without cover crops. There are thus six combinations [rotation X fertilisation X cover crop] which are simulated over 30 years for this PCU, and for two soil types (for a total 12 simulations over 30 years for this PCU).

The coherence of the simulation results in terms of yields and quantity of aboveground biomass at harvest was verified through comparison to agricultural statistics on annual yields at the departmental level. Following this analysis, out 32,318 simulations of "current" cropping systems (= combinations of [soil type X rotation X fertilisation X cover crop]), 30,580 cropping system simulations were retained for analysis.

Analytical strategies for the simulation results

For each output variable from the model, the value calculated at the level of the PCU (allowing for the creation of maps) corresponds to an average of obtained values for each combination [soil type X rotation X fertilisation X cover crop] weighted according to the importance of each of these four factors within the PCU land area. The variables were estimated on a per-day basis and then aggregated on a yearly basis in order to calculate the ES indicators either as annual averages over 30 years or as the difference between the initial state (value at the beginning of the simulations) and the final state (value obtained for year 30 of the simulation), depending on the needs of the analysis.

In order to test the effect of cover crops on the level of ES supply, alternative simulations "without cover crops" (other practices unchanged with respect to the reference simulations) were conducted for the PCU located in Nitrate Vulnerable Zones (according to the Nitrate Directive), where installing cover crops in between primary crops is mandatory.

Results from the STICS model (field crops)

Results were examined with respect to three variables: i) the texture of the soil surface horizon; ii) the initial OM condition (amount of soil OM at the beginning of the simulations), which determines the initial level of supply of mineral N through mineralisation for a given climate; and iii) the length of the rotation.

The average quantity of N supplied by the ecosystem during the period of crop growth (from sowing to harvest) varied from 42 to 224 kg N/ha (average of 93 kg N/ha) (Figure 2-3). The total quantity of mineral N available to the commercial crop (including the amount of mineral N present in the soil at the time of sowing) followed the same general pattern of spatial distribution, but its average value at the national level was 143 kg N/ha.

The highest values are found for the most part in Aquitaine, Alsace, and Brittany, in parts of the Parisian basin and along the River Saône. The lowest levels are found in the mid-Garonne Valley, around the perimeter of the Parisian basin, in Lorraine, in Limagne, and in the southern Rhône Valley.

For both indicators, clayey soils are associated with much lower average values than other types of soils. It was not possible to investigate this finding further in EFESE-AE, but doing so would involve examining the observed relationships between soil type and crop type.

In addition, the results showed indicator values that were both higher and more variable for monocultures than for multi-crop, multi-year rotations. One can hypothesise that for continuous wheat or continuous maize, both often heavily fertilised, crop residues (straw or stover) left in the field provide a significant amount of mineralised biomass for the following year's crop. This result shows an indirect effect of N fertilisation.

Beyond the range of values for these two indicators, the main difference between them lies in their sensitivity to initial soil OM levels. In general, the amount of mineralised N tends to be higher when initial OM levels are high. The median value is situated around

75 kg N/ha for soils with the lowest initial OM levels, and 95 kg N/ha for soils with the highest initial OM levels. This effect, which remains to be confirmed by statistical analysis, disappears when one accounts for the amount of mineral N in the soil at the time of sowing. This suggests that the direct effect of fertilisation (the effect of inputs in year $n - 1$ on the amount mineral N in the soil at the time of sowing for year n) is stronger than the effect of mineralisation of residual biomass.

Figure 2-3. Total mineral N supplied by the ecosystem (in kg N/ha) as estimated by STICS for cropping systems managed with current observed agricultural practices

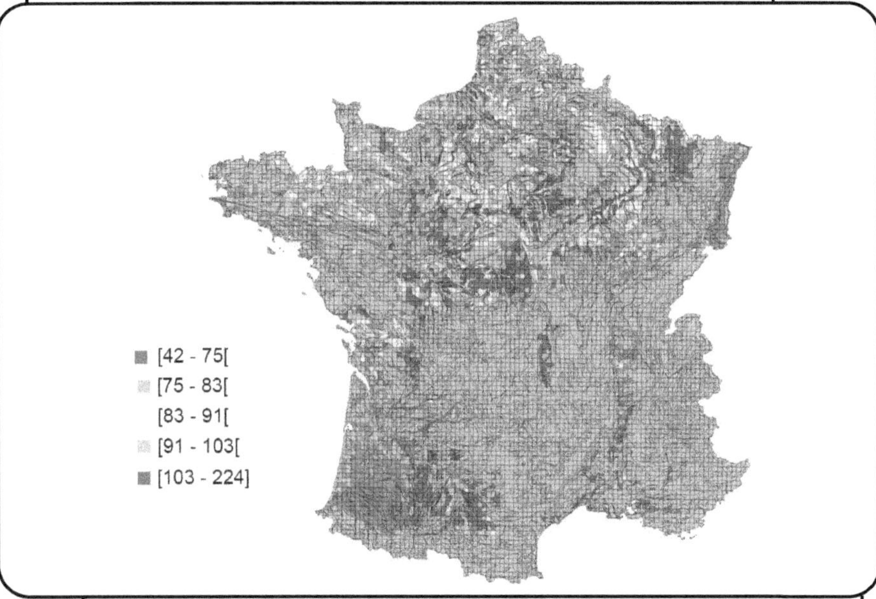

Spatial resolution: PCU; PCU in gray (including Corsica): no "field crops" simulations; PCU in white: excluded from the analysis; Value classes are expressed in quintiles.

Effects of cover crops on the level of ES supply

For PCU located in Nitrate Vulnerable Zones, the simulation results showed that the presence of a cover crop increases the amount of N supplied by the ecosystem to the commercial crop. This finding confirms the hypothesis that the mineralisation of cover crop residues increases the quantity of N available in the soil. The effect is modest, however, with an average difference of 5 kg N/ha between the median values of the indicator simulated "with" and "without" the use of cover crops, and a considerable overlap between the two ranges of values. Furthermore, the presence of a cover crop does not appear to have an effect on the amount of N in the commercial crop at the time of harvest.

Perspectives for improvement

The simulation rules adopted here represent a methodological advance relative to classic studies relying on single-year or multi-year simulations in which the condition of the soil is "reinitialised" each year or for a sequence of years. Continuing the simulations over a 30-year timespan makes it possible to account for the effects of preceding crops (both commercial crops and cover crops) – in particular, the amount of N supplied by crop residues. On the other hand, the analytical protocol could be improved further, notably by accounting separately and specifically for winter crops vs. spring crops, since the dynamics of N mineralisation and symbiotic fixation are strongly influenced by seasonal climate variations.

The two indicators used to describe the level of ES supply are dependent on external N inputs. In the simulation design, the nature and amount of nitrogenous fertilisation were estimated at the regional level based on data with respect to prevailing practices. To avoid overestimating fertiliser use in marginal areas with significantly lower yield potentials relative to high-production areas within a given region, one should develop a method to adjust input levels in accordance with yield potential. For example, one could perform an initial set of simulation scenarios, then estimate yield potentials by PCU, adjust fertilisation rates where they are clearly overestimated and then redo the simulations using these adjusted fertilisation rates.

▌ Storage and return of water by the ecosystem

Ecosystem water resources are typically classed in two categories: "green" water – precipitation stored in the soil and returned to the atmosphere through plant transpiration and soil evaporation – and "blue" water – water in lakes, rivers, oceans, and aquifers.

An ecosystem's capacity to store and return water provides two interdependent but distinct ES due to the nature of the benefits to the farmer and society as a whole. The first ES, "storage and return of water to crop plants," directly benefits the agricultural ecosystem manager. The benefit the farmer receives corresponds to the quantity of water he or she would have to supply through irrigation to obtain the same level of crop output in the absence of the ES. The second ES, "storage and return of blue water," directly benefits society as a whole, which makes use of this water resource for a variety of purposes: agricultural, industrial, domestic, recreational, and cultural. Although the latter does not constitute an input SE, its presentation is integrated into this section.

Biophysical determinants and external factors
Biophysical determinants

ES linked to the storage and return of water depend for the most part on processes of water flow – under the influence of gravity (percolation), through runoff at the soil surface, and through lateral movement within soil layers – and processes of evapotranspiration. These processes depend on the nature of the plant cover, on the dynamics of its growing cycle, and on soil properties and characteristics (soil moisture levels and the texture and percentage of organic matter, which determine soil porosity). The condition of the soil surface and the degree of vegetative cover determine the division between infiltration and runoff. Two poorly understood hydric processes contribute to the movement of water upwards in the soil and the plants: nocturnal exudations of water by plant roots, and upward capillary movement. The latter is more significant than the former, and may be responsible for 30 to 60% of crop transpiration in areas with a shallow water table.

It is important to note that there is often an intrinsic biophysical antagonism between percolation through the soil and evapotranspiration at the soil surface. As a general rule, it is observed that when stored soil moisture is being evapotranspired by plants to form biomass, the amount of water percolating (or draining) through the soil is small; conversely, the more water percolates through the soil when less water is being evapotranspired by plants.

External factors

The capacity of the agricultural ecosystem to store and return water is affected by climate, in particular by the amount and distribution of precipitation throughout the year. Precipitation represents the entry of water into the ecosystem. It has been shown that the relationship between the amount of water percolating through the soil and the amount evapotranspired by plants is strongly determined by climate: generally speaking, in a dry year, the percentage of rain water evapotranspired is distinctly higher than the percentage percolating through the soil; whereas the reverse is true in wet years.

These two ES are also strongly dependent on agricultural practices, notably fertilisation, crop residue management (presence of a surface mulch), soil tillage, and irrigation. With respect to the latter, we can note that the soil's capacity to store and return water determines the efficiency of various types of irrigation practices (as a complement to ES supplied to the farmer) in meeting crop water requirements (see Chapter 3).

Level of ES provision

The approach adopted in EFESE-AE to examine ES relating to water flow is radically different from that adopted by the CICES (and thus by other work following the CICES model). The MAES program, for example, considers four ES for water supply, distinguishing the type of water bodies (surface vs. underground) and their use (food vs. other uses). The indicators he proposes turn out to be inappropriate here.

Other studies seeking to quantify water flow concentrate on the distinction between green water and blue water. By definition, the indicators proposed to quantify the movement of green water are not useful in quantifying water stored and returned by the ecosystem to crop plants, since they do not distinguish between transpiration and evaporation. Moreover, the quantity returned by the ecosystem in the form of blue water – that is, moving through percolation, run-off, or lateral hypodermic drainage – is typically described as the water yield, defined as the difference between annual total precipitation and the annual quantity of water evapotranspired.

Assessment method

The dynamic simulation framework developed in EFESE-AE (see Box 2-1) was used to estimate the different components of the water balance, according to which the variation ΔS of the stock of water available to plants in the soil (available water reserve) is calculated as follows:

$$\Delta S = \text{Precipitation} + \text{Irrigation} - \text{Evaporation} - \text{Transpiration} - \text{Runoff} - \text{Drainage}$$

Two indicators were defined:

- one indicator for the ES "storage and return of water to crop plants": the quantity of water transpired by the commercial crop between planting and harvest;
- one indicator for the ES "storage and return of blue water": the water yield calculated for the year, as running from September 1 in year n to August 31 in year $n+1$.

The two indicators of annual average ES levels were calculated based on simulations at the daily level so as to take into account water flow interactions during critical periods – such as periods of abundant rainfall, when rainfall effects on percolation are strongly dependent on the type and developmental stage of the crop.

Several sets of simulations were performed:

- Simulations "without irrigation", including for maize crops that are typically irrigated, were performed in order to quantify the level of ES provided by the agricultural ecosystem by excluding external water inputs. Other agricultural practices were not modified, and represented currently prevailing management methods for agroecosystems.
- Alternative simulations were performed for certain PCU in order to test the effect of two agricultural practices on the level of ES supply:
  - alternative "no cover crop" simulations (other practices unchanged relative to the reference simulations) for PCU located in Nitrate Vulnerable Zones (according to the Nitrate Directive), where use of cover crops between primary crops is mandatory;
  - alternative "with irrigation" simulations (other practices unchanged relative to the reference simulations) for PCU containing maize crops that are normally irrigated.

Results

As before, only the results obtained for major field crop ecosystems are presented. Results were interpreted with regard to three variables: i) the maximum available water reserve (MAWR), or the maximum quantity of water the soil can retain and return to roots for plant life; ii) the length of the rotation; and iii) the climate type based on average precipitation and annual observed temperatures for mainland France.

Storage and return of water to crop plants

The first simulation scenario – "no irrigation" – made it possible to estimate the level of ES effectively supplied by the ecosystem, which is to say effectively exploited by the agricultural ecosystem manager for the production of biomass. Across all PCU considered, the average quantity of water transpired annually by the commercial crop varied from 63 to 295 mm (average of 153 mm – Figure 2-4).

Figure 2-4. Estimated average annual transpiration (in mm) from commercial crops in cropping systems without irrigation (including for PCU where crops are typically irrigated)

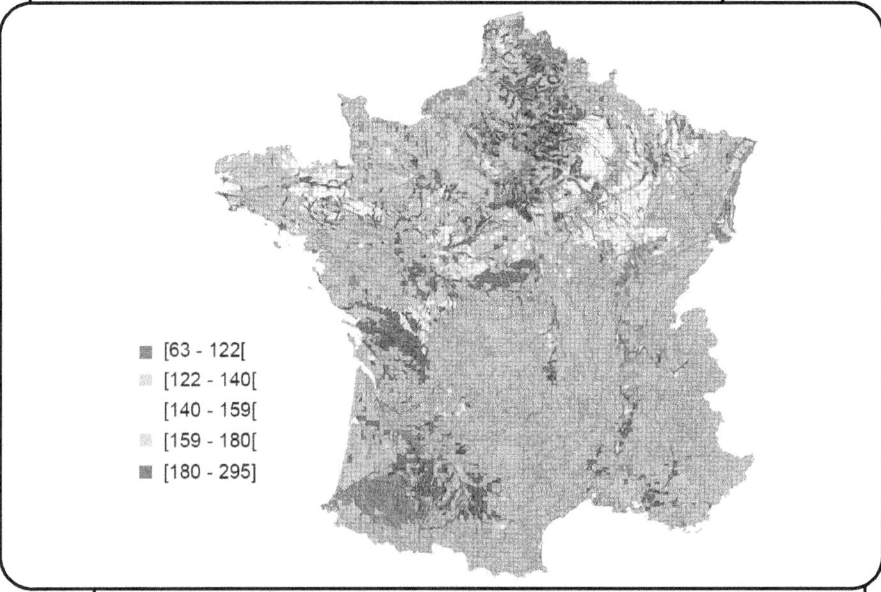

Spatial resolution: PCU; PCU in gray (including Corsica): no "field crops" simulation; PCU in white: excluded from the analysis; Value classes correspond to quintiles.

The highest levels of ES are found in the foothills of the Atlantic Pyrenees, in the center and northern part of the Parisian basin (which also shows locally low values), in the Alsatian Plain, and in the Saône Valley. At the other extreme, the lowest levels are found in the mid-Garonne Valley (except immediately along the river), in Poitou-Charentes, to the west of the Parisian basin (Sarthe, Indre-et-Loire, Vienne), in Berry and in Limagne. Low values are also found more locally in the Rhône Valley, in Brittany, and in Lorraine.

Transpiration varies essentially as a function of the MAWR. On average, transpiration is higher where MAWR is higher. The indicator values are evenly divided between the four MAWR classes defined at the outset.

Storage and return of blue water

Annual average water yield varies from 55 to 1119 mm across all PCU, with an average of 315 mm (Figure 2-5). Flows of blue water are thus on average two times greater than flows of water returned to crop plants.

The highest values correspond to PCU in the Landes region, in Brittany, on the periphery of the Parisian basin (from the Channel coast in the west to the Barrois and Langrois plateaus in the east), and in Rhône-Alps. At the other extreme, the lowest values are found in the mid-Garonne Valley, in Touraine, in the heart of the Parisian basin, in Flanders, on the Alsatian Plain, in Limagne, and in the southern Rhône Valley.

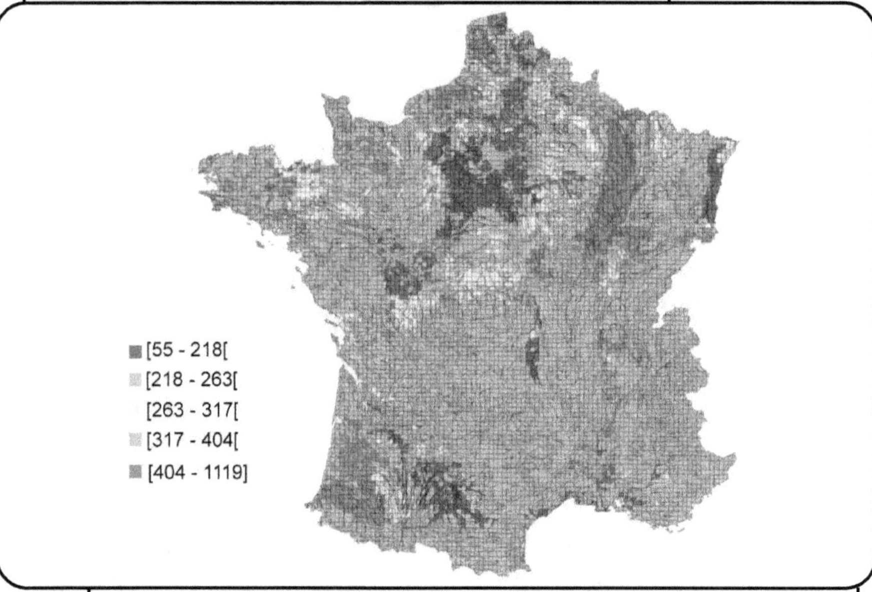

Figure 2-5. Annual average estimated water yield (in mm) for cropping systems without irrigation (including for PCU where crops are typically irrigated)

Spatial resolution: PCU; PCU in gray (including Corsica): no "field crops" simulation; PCU in white: excluded from the analysis; Value classes correspond to quintiles.

Water yield varies primarily as a function of climate type, and in a strongly contrasting fashion. The typical climate of "Southwestern Basin," is characterised by low annual precipitation levels and is associated with the lowest water yields (200 mm on average). Conversely, the "Mountainous" climate type is associated with the highest values (500 mm on average). This climate type is characterised by high precipitation amounts, low temperatures, and thin soils, all factors that favor water percolation through the soil profile.

Effect of irrigation on water flow

The alternative simulations "irrigated maize" clearly demonstrate the positive effect of irrigation on transpiration in the commercial crop, notably for intermediate MAWC (40 to 120 mm). This effect most likely involves simulation units in the Landes region, which has considerable land area in continuous maize. Irrigation does not seem to have an effect on water yield, however.

It is worth qualifying these results with regard to the way in which irrigation is modeled in these simulations, however. In the absence of empirical data on actual irrigation practices, water inputs are calculated by the STICS model based on the principle of limiting crop water stress (irrigation to supply 85% of crop water requirements). By definition, this tends to maximise the efficiency of water inputs vis-à-vis crop needs, and as a result tends to strongly limit blue water.

Effect of cover crops on ES levels

The alternative simulations "no cover crop in PCU in Nitrate Vulnerable Zones," do not show an effect of the cover crop on commercial crop transpiration rates, nor on annual water yield. More detailed analyses should be done to confirm these results. In addition, new approaches to cover crop management – e.g., longer coverage periods – could be tested, which could lead to a revision of these results.

Perspectives for improvement

In this assessment, the water balance was modeled using a "reservoir"-type approach, assuming that rain and irrigation to be the only water inputs into the ecosystem, and neglecting preferential vertical drainage and upward capillary action. Again for the purposes of simplification, simulations assumed that all fields had a zero slope and thus that there would be no lateral water movement (in the form of runoff or hypodermic drainage). This modeling decision made it possible to supply information on water movement for level fields within the different PCU of France. It would however be poorly suited to an analysis attempted at the watershed level, given the relevant geomorphology.

In future assessments, a differentiated analysis of indicators by crop type should be undertaken (percentage of spring crops vs. summer crops, or maize crops vs. other crops). The climate typology could also be refined so as to account for inter-seasonal variability and thus allow for a more precise interpretation of the chosen indicators. Again, a goal here would be to identify the effects of winter crops vs. spring crops.

▌ Soil stabilisation and erosion control

Erosion corresponds to the movement of materials transported away from the soil, mainly by water and wind. In agricultural ecosystems, erosion results in a loss of soil from the surface soil layers, which are richest in organic matter. The ecological processes that contribute to holding soil constituents and surface sediments in place can thus be defined as an ES of "soil stabilisation and erosion control."

The agricultural ecosystem manager benefits from this ES in terms of the preservation of soil capital and the maintenance of the agronomic potential of the production setting. Stabilisation of the arable soil layer determines the level of nutrient supply to crop plants and the storage and return of water to the ecosystem, thus helping to reduce the amount of fertilisers and water that must be supplied to maintain a given production level.

Beyond the agricultural ecosystem context, this ES also provides direct benefits to society, helping to limit catastrophic events such as mudslides and improving water quality through the reduction of waterborne soil particles.

Biophysical determinants and external factors

Biophysical determinants

The plant cover and, to a lesser extent, the mineral and organic state of the soil are the two main biophysical determinants of the level of ES supply. Erosion by physical agents other than water (wind, avalanche) were not considered in EFESE-AE.

Several research studies have suggested that erosion and runoff decrease exponentially with the degree of plant cover in both space and time. Vegetation intercepts rainfall, reducing its capacity to disaggregate surface sediments; it also favors soil infiltration along plant root systems. Finally, the more plant cover there is and the higher its foliar index, the more evapotranspiration there will be, to the detriment of lateral water flow and percolation. It is thus particularly important to consider spatial and temporal variations in plant cover.

The mineral and organic state of the soil (texture, organic matter levels, permeability, etc.) determine its structural stability, which in turn determine two key properties: sensitivity to soil crusting and soil erodibility. Sensitivity to soil crusting corresponds to the soil's propensity to form a superficial clogging crust by destructuring the surface layers due to low structural stability.[22] This process limits the soil's capacity for water infiltration. Soil erodibility determines the soil's susceptibility to disaggregation and the removal of material by rainwater.

Finally, the risks of runoff are naturally higher on sloping fields.

External factors

Certain agricultural practices act as exogenous factors affecting the level of ES supply. Studies have shown that an intensification of soil tillage has a tendency to erode soil from convex points and redistribute soil and organic matter at concave points. These phenomena will obviously be accentuated by deep plowing or by plowing parallel to the direction of the slope. Irrigation practices can also affect this ES through their influence on the degree of plant cover and in some cases through an impact on the disaggregation of soil materials.

Finally, for a given agricultural ecosystem (soil, vegetative cover, topography) with a given potential for soil stabilisation and erosion control, precipitation regimes determine the level of erosion and/or runoff. Some precipitation regimes have a greater impact on the level of ES supply by agricultural ecosystems, depending on i) the percentage of rain falling during periods of exposed soil and ii) the frequency of extreme rain events. All other things being equal, in situations where the available water reserve is low, the risk of runoff will be higher since the soil will reach saturation more quickly during rain events.

---

22. Resistance of soil structure to degradation agents.

Level of ES provision

ES assessment is usually undertaken through an estimate of the difference in erosion rates between the situation to be assessed and a reference situation. A large number of empirical models have been developed to simulate soil erosion rates at the European level by combining information layers for pedological, climatic, and agroecological data. Thus, a RUSLE (Revised Universal Soil Loss Equation) model has been used to quantify the level of ES supply for the MAES program. The MESALES[23] model, developed by Inra, simulates an erosion risk that can be converted into an erosion rate using a correspondence table. By comparison with the RUSLE model, MESALES assigns greater importance to certain pedological factors such as the soil's susceptibility to rain splash. More mechanistic models describing the physical processes involved in erosion, such as PESERA (PAN-European Soil Erosion Risk Assessment), are currently being developed, but no method for the generalised spatial validation of this model appears to have been completed to date.

The principal strength of the RUSLE and MESALES models is that they are based on a very flexible modeling framework, requiring few parameters and making use of databases that are already available at the European level.

Assessment method

The MESALES model was used in EFESE-AE to quantify the ES for "soil stabilisation and erosion control" by comparing the erosion rates for the "current" situation (current plant cover and susceptibility to soil crusting and erodibility, represented using available databases) and for a reference situation. The level of ES effectively supplied by a given agricultural ecosystem – corresponding to the quantity of soil stabilised by the ecosystem, considering its topographical configuration and current plant cover – was estimated through a calculation of the difference in erosion rates between and the "current" situation a "bare soil" reference situation.

To assist in the interpretation of this indicator, a "relative" ES level (compared to a hypothetical maximum) was also estimated by calculating the ratio between the "current situation" ES level and the ES level of a "permanent cover" situation assumed to supply the highest possible level of ES. This second indicator expresses the percentage of the maximum ES currently supplied by the agricultural ecosystem.

Major improvements were made in EFESE-AE with respect to the implementation of MESALES at the level of France as a whole. The most significant of these was to refine the description of the plant cover both across space (proportion of surface area covered) and time (by describing more precisely the stage of development of the cover) from the analysis of spatial remote sensing data.

---

23. For *Modèle d'évaluation spatiale de l'aléa érosion des sols* (Spatial assessment model of soil erosion hazard).

Results

In relatively flat parts of the country (Landes, Beauce, the Alsatian Plain), the annual ES level was nearly zero, since the erosion rate for bare soil is very low in these topographical conditions (Figure 2-6).

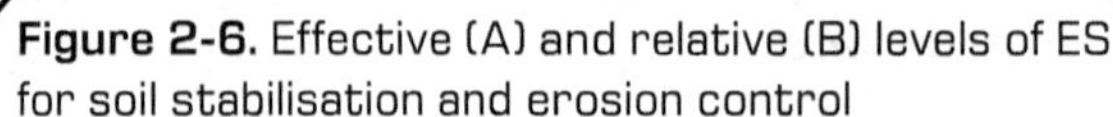

Figure 2-6. Effective (A) and relative (B) levels of ES for soil stabilisation and erosion control

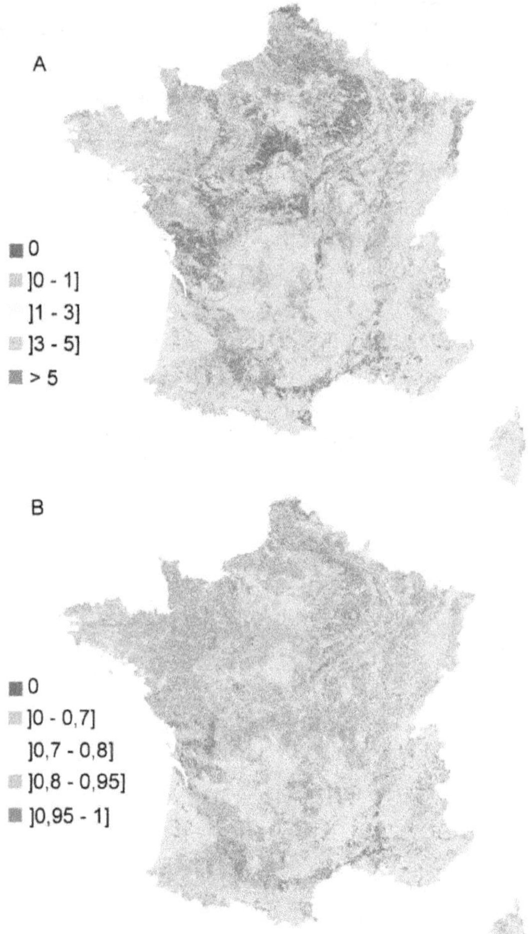

A: Effective ES level, in t/ha/yr, calculated for each 100 m square cell;
B: Relative ES level, no units, calculated for each 100 m square cell;
Gray pixels: no agricultural ecosystem.

The highest ES levels were observed in the major grass-growing areas (Brittany, Lower Normandy, the Massif Central, the Alps, the Jura, etc.). In these areas erosion rates are very high without permanent vegetative cover. Conversely, zones with low absolute and relative ES levels correspond to the major crop-growing regions (wheat in 30% of cases, maize in 10%, oilseed rape in 5%, sunflower in 5%, etc.) or to perennial crops (20%) in areas with high maximum ES levels: north and east of the Parisian basin, the foothills of the Pyrenees, and some parts of Brittany, Midi-Pyrenees, Languedoc, the area around Lyon, etc.

The results did not allow for the identification an effect of the type of crop on the level of ES supply. Statistical analyses would be needed to examine the correlation between ES levels and the presence of specific practices or types of plant cover assumed to be favorable to expression of the ES. Indeed, at first glance, the (qualitative) examination of the mapping seems to suggest that areas where grasslands are three times more present than arable crops (e.g., the Creuse) have lower ES levels than areas where grassland is less present (for example, in the foothills of the Atlantic Pyrenees).

Examining the results by season shows that in most areas, the highest ES levels are reached in the winter, and secondarily in the fall – the two seasons with the highest precipitation amounts. This is especially true for Brittany and Lower Normandy, where it can probably be attributed to significant levels of soil cover for arable lands at these times of year. In other regions, however, such as the southwest (the Adour basin) or the area around Lyon, ES levels tend to be higher in the summer than in other seasons. These areas probably have less plant cover during the rest of year (areas dominated by spring crops without the use of cover crops).

Perspectives for improvement

The principal weakness of the MESALES and RUSLE models relates to uncertainty associated with the entry variables, notably those describing soil properties and modes of land use. One way to improve the quality of the assessment, therefore, would be to identify more precise data relating to i) land use, so as to account for the spatial distribution of semi-natural elements, and ii) intra-seasonal and inter-annual dynamics of plant cover, so as to study the effects of cover crops on ES levels. The assessment could also be extended to include an analysis of the effects of climate change and changes in land use over a decade or more.

 **Biological regulation**

ES of "biological" regulation are defined here as processes that regulate biotic factors of agricultural production. This category includes ES for crop pollination and ES for the regulation of pest species. With regard to the latter, two major types of processes may be distinguished: i) regulation linked to the aboveground faunal species, also referred to as the "natural" flying and plant-level biodiversity (so-called "top-down" interactions between pests and their natural enemies); and ii) regulations linked to the spatio-temporal configuration of plant biodiversity, including both planned (cultivated) and associated (weed species and semi-natural habitats within the field and its immediate vicinity) plant populations (so-called "bottom-up" interactions between pests and plants).

▌ Pollination of cultivated species

Pollination is the process of pollen transfer from the male to the female reproductive parts of the flowers of Angiosperms. The analysis focused on pollination involving living organisms, that is, pollination performed by animal vectors (as opposed to wind pollination). Approximately two-thirds of crop species rely on pollination of this type, primarily fruit and vegetable crops, but also some oilseed crops.

Since pollination is a biophysical determinant in the production of plant goods, farmers receive direct benefits from this ES in the form of increased crop outputs. In situations where a potential pollination deficit would lead the farmer to implement alternative pollination strategies (e.g., beehive rental, manual pollination), ES for crop pollination can provide a benefit in the form of avoided costs.

Biophysical determinants and external factors
Biophysical determinants

The principal biophysical determinants involved in the supply of this ES are i) the structure of pollinator communities (abundance, diversity), and ii) the characteristics of wild and cultivated plants that require pollination.

In temperate regions, pollinator communities are primarily composed of insects. The analysis presented below thus focuses on insect pollination. The structure of pollinator communities depends on the composition and configuration of the landscape matrix surrounding the agricultural ecosystem, usually consisting primarily of semi-natural habitats adjacent to fields (woods, tree rows, hedgerows, road margins, etc.) that offer habitat, nesting areas, and food resources (wild plants) for pollinators.

The physiology and morphology of plants determine their dependence on pollinators and their degree of specialisation with respect to pollinator taxa. The spatio-temporal diversity of managed covers, as defined by the farmer's cropping sequence, will determine the availability of alternate floral resources in between crop flowering times.

External factors

Certain agricultural practices can be considered as external factors likely to influence the intensity and the efficacy of crop pollination through their effects on these biophysical determinants. Numerous studies have sought to examine the negative effects of certain agricultural practices on pollinator diversity and/or abundance, often through a comparison of "conventional" vs. organic agricultural systems. Broadly speaking, the most important agricultural practices affecting ES for crop pollination are soil tillage and pesticide use.

Finally, climate change can lead to changes the dynamics of plant development (advancement of flowering, decline in leaf fall, etc.) and/or in the distribution areas of pollinators. These modifications could result in spatial, temporal or functional disjunctions between plant and pollinator species. These impacts are difficult to observe, however, and their implications for crop pollination have not been clearly established.

Level of ES provision

Quantifying insect pollination levels can be done through an estimation of the effects of pollination on seed output for a given plant, taking into consideration any limitations on plant nutrient resources likewise necessary to seed development. Experimental protocols for this type of quantification have not been developed at the national level. A large number of studies have proposed an indirect evaluation of this ES, however. Existing indicators focus primarily on the relationship between the availability of semi-natural habitats and the composition of pollinator communities. For example, the indicator developed by Zulian *et al.* (2013),[24] as part of the MAES program, calculates a relative pollination potential by combining: i) an estimate of the capacity of various landscape elements to supply food resources and nesting sites, ii) a maximum travel distance (for solitary bees), and iii) an activity index for weather-dependent pollinators. This indicator thus provides information with respect to the potential distribution of pollinator habitat, and the area served by pollinators based on this habitat.

Assessment method

In EFESE-AE, three indicators were used as proxies for the level of supply of this ES.

An initial attempt was made using the indicator for relative pollination potential as employed by the MAES program. This does not include direct information on the composition and abundance of pollinator communities, however.

To complement these results, an indicator for pollinator species richness was developed by extrapolating to the level of mainland France the observational data collected for the participatory program SPIPoll (Photographic Monitoring of Pollinator Insects).[25] Four groups of pollinators were considered: Hymenoptera, Diptera, Lepidoptera, and Coleoptera.

Finally, a pollination index, a more direct proxy for the level of ES supply, was developed and calculated for each Department of metropolitan France, based on an approach

---

24. Zulian, G., Maes, J. & Paracchini, M. 2013. Linking land cover data and crop yields for mapping and assessment of pollination services in Europe. Land 2: 472–492.
25. www.spipoll.org

developed by recent research. The underlying assumption is that a pollination deficit will result in a yield deficit, and that this deficit will be greater for crops that are more strongly dependent on insect pollination. This indicator was calculated using statistical yield data and coefficients of crop pollinator dependence available in the scientific literature.

The first two indicators provide information as to the level of ES potentially supplied by agricultural ecosystems. The third – the robustness of which remains to be determined – is intended to supply information as to the level of ES effectively supplied to the farmer.

Results

Figure 2-7 presents the results obtained for the three indicators described above.

Roughly speaking, the values of the three indicators increase along a North-South gradient. In statistical terms, however, this positive correlation is only moderate. A stronger relationship is observed between the indicator of relative pollination potential (indicator A in Figure 2-7) and the pollination index based on yields (C). Since these are based on two completely independent datasets, one is led to believe that the apparent North-South gradient is a robust result. In the literature, the existence of this gradient is attributed to a latitudinal variation in temperatures that favors bee activity at low latitudes. In France, this gradient is probably also linked to landscape configuration, with a weaker presence of semi-natural habitats in northern France (the greater Parisian basin). We can note too that the distribution of pollinator morpho-species (B) along this gradient is a recognised, multi-factorial result in the literature, and has likewise been observed for nearly all the taxonomic groups.

The more remarkable difference among the three indicators is the low level of pollinator diversity in the Mediterranean basin (B). These low values are due in large part to variations in the richness of Diptera, which represent a significant percentage of sampled morpho-species in the SPIPoll program and which are not very diverse in the Mediterranean region because the climate is too hot for them. Indicator A is valid for the "solitary bee" model, and thus does not provide reliable predictions for Diptera. The divergence between indicators B and C could be explained by a reduced dependence on pollination by Diptera in the Mediterranean basin, but this hypothesis remains to be tested given that recent research suggests a major role for Diptera within ES for "crop pollination" in general.

Moreover, none of the three indicators has been validated empirically, and their interpretation should be undertaken with caution. Indicator A is based exclusively on the example of the solitary bee, whereas other groups of pollinators contribute in a non-negligeable manner to the ES. It thus probably underestimates the level of ES supply. Broadening this indicator to other groups of insects would require data on the ecology of these groups, which are currently scarce but are in the process of being acquired (large-scale monitoring programs such as SPIPoll). Indicator B measures pollinator species richness, whereas several studies have shown that pollination

Figure 2-7. Three indicators of the level of the ES for "crop pollination" in agricultural ecosystems: relative pollination potential (A), pollinator species richness (B), pollination index (C)

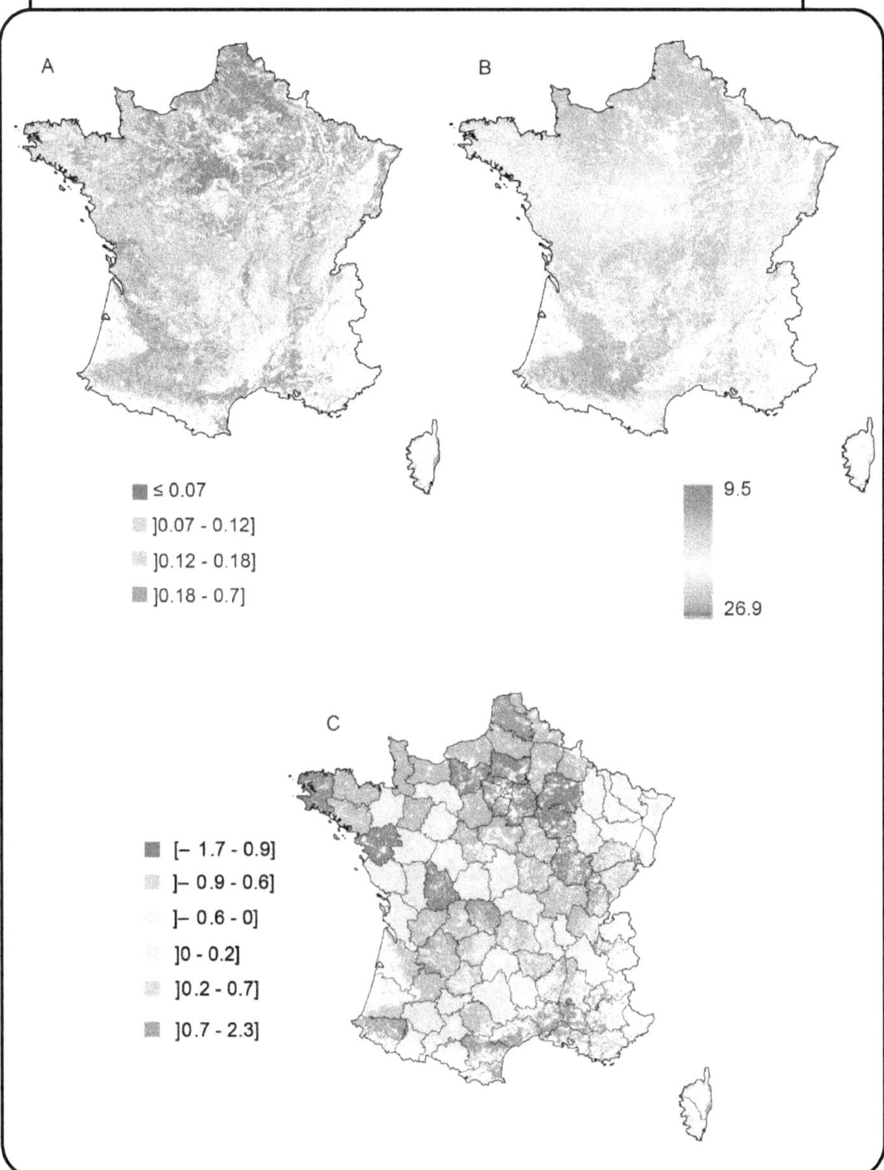

A: Relative pollination potential, no units; calculated for each 100m square cell (Zulian *et al.*, 2013); B: Pollinator "species" richness, number of "species-types" for each 100 m square cell; C: Pollination index, no units; calculated for each administrative Department.

efficacy depends more on functional diversity than on species richness *per se*. This indicator is thus only relevant if the morpho-species correspond at least partly to functional groups, a hypothesis that remains to be verified. Finally, part of the gradient of values obtained by calculating the indicator C may be explained by a local adaptation of cultivated species to pedoclimatic conditions, correlated with their dependence on pollinators. In the Mediterranean basin, for example, pedoclimatic conditions are generally unfavorable to major field crops that are not pollinator dependent, such as wheat, and more favorable to fruit trees, which are highly pollinator dependent. This phenomenon may lead to an underestimate of the real level of the ES for "crop pollination," by an amount that is difficult to quantify.

A first correlation analysis between the different indicators suggests that at least part of the North-South gradient is due to variations in the level of the ES for "crop pollination." In addition to the limitations previously cited, improving the prediction of ES levels by these indicators will involve improving the quality of the entry data, including: (i) a better accounting of small semi-natural features (hedges, forest patches) so as to improve the relevance of indicator A at the local level; (ii) enriching the dataset on pollinator species so as to eliminate the over-representation of urban ecosystems and improve the spatial extrapolation method on which indicator B is based; and (iii) collecting data on crop yields at a finer level of resolution (e.g., the SAR), since the departmental level as currently used strongly penalises the quality of the information supplied by indicator C.

▌ Conservation pest control (weeds, insect pests)

For the agricultural ecosystem manager, biological control offers an alternative to using pesticides to curb pest and weed pressures on crops and thereby protect crop yields. Biological control can take two forms: i) the introduction of crop auxiliary species (biological control by acclimatisation), or ii) the management of ecosystem entities and processes (biological control by conservation). Within the framework of ecosystem services assessment, we will focus here on the second of these.

Biological control by conservation corresponds to natural processes of pest regulation, including all crop enemies. These can be divided into three major groups: pathogenic agents (plant diseases), pests that eat crops, and weeds that compete with crops. Despite the importance of plant pathogens as a factor in crop production, it was not possible to include disease regulation within the scope of EFESE-AE.

ES for the regulation of weeds and the regulation of crop pests directly benefit the farmer through their contribution to the maintenance of crop yields and reductions in the use of crop protection products (some of which have known negative effects on farmers' health) and/or mechanical cultivation requirements.

Biophysical determinants and external factors

Biophysical determinants

The spatio-temporal configuration of plant cover is a major determinant of the weed seed bank and pest insect populations (bottom-up regulation). Numerous studies have shown that changes in the spatio-temporal distribution of planned biodiversity (e.g., cover crop use, changes in planting dates, adjustment of row spacings) serve as an effective management lever for weed pressure. Crop rotations in particular have a significant effect on the abundance of the soil weed seed bank and thus on floral development over the long term. With regard to pest species regulation, it is crop diversity within the parcel that has been shown to have the greatest effect. Other studies have demonstrated the importance of weeds as habitat features for associated biodiversity and thus as key elements in biological regulation.

In addition, the regulation of both weed seeds and pest insects rely on processes of parasitism and predation involving other predatory taxa (top-down regulation). Within agricultural ecosystems, it has been shown that carabids (and to a lesser extent certain bird species) are predator taxa for weed seeds. The predators and parasites of crop pest insects that are naturally present in agricultural landscapes are extremely diverse, including arthropods (ladybugs, lacewings, carabids, spiders, etc.), birds and even mammals (bats). These predator taxa move within the agricultural ecosystem and its immediate surrounding environment, which supply them with food reserves, habitat, overwintering sites, etc. As a result, the composition (particularly the percentage of semi-natural areas) and configuration (proximity to agricultural fields) of semi-natural habitats have an impact on communities of weed seed consumers.

More generally, the principal biophysical determinants of ES for the natural regulation of weeds and crop pests are:
  • the characteristics of the pest species: abundance of the weed seed bank in the field and the structure of communities of plant-eating arthropods;
  • the configuration of the planned and associated plant diversity within the field over time (crop rotation sequence) and in space (diversity, density, layout);
  • the structure of communities of crop auxiliary taxa: abundance and diversity of weed seed predators and natural enemies of pests in the field;
  • the composition and configuration of the landscape matrix surrounding the field, in particular the presence of semi-natural habitats.

External factors

In addition to the overall influence of the climate on the biophysical determinants mentioned above, the use of crop protection products and soil tillage practices are the principal external factors that influence the level of ES for biological control. Weed control and soil tillage practices exert selection pressure on weed species and thus shape the composition of the weed seed bank over time. These practices also have direct and indirect effects on crop auxiliary species (destruction of nests and/or individuals, reduction in food resources for these taxa, ecotoxicity).

Level of ES provision

Research conducted in the United Kingdom by Bohan *et al.* (2011)[26] has established a model predicting weed seed bank abundance as a function of crop rotation type. This method makes it possible to quantify the impact of the ES for weed seed regulation by means of crop rotations, a major component of the ES for the regulation of weed seeds. The method has yet to be implemented for an entire country, however.

The principal methods for measuring the level of biological control by natural predators in the field are: i) monitoring of predator populations present in agricultural fields and the measurement of weed seed "rain" using *in situ* trapping methods; ii) measurements of yields and damage to crops in the field to produce an estimate of pest population impacts; and iii) measurement of predation levels using tools such as seed cards. Experimental protocols for implementing these types of measurements exist, but have never been attempted for France as a whole – only local databases exist. Several studies have shown that certain landscape characteristics can serve as a proxy for the diversity and/or abundance of some types of crop auxiliary species, or even for potential levels of pest regulation, suggesting the possibility of predicting the abundance of weed seed predators and of natural enemies of crop pests based on a knowledge of landscape composition and configuration. No method of this type has been developed at the level of France as a whole, however.

Assessment method
Regulation of weed seeds through the use of crop rotations

One indicator of the effect of crop rotations on weed seed bank abundance was constructed using the model established by Bohan *et al.* (2011) for data from the United Kingdom. The model was applied to crop rotations observed in France. This initial indicator made it possible to estimate weed seed bank abundance based on current cropping systems, and thus ES levels potentially supplied by agricultural ecosystems.

Biological control by natural predators

Two methodologies for quantifying the potential level of supply of these ES were developed using previous research on the relationship between landscape composition and ES levels for biological regulation in major field crops. Implementation of these methodologies at the national (French) level sought to illustrate their potential for the assessment of ES of biological control by conservation, considering the current state of research and available datasets. For each methodology, the indicator adopted for

---

26. Bohan, D.A., Powers, S.J., Champion, G.T., Haughton, A.J., Hawes, C., Squire, G.R., Cussans, J. & Mertens, S.K. (2011) Modelling rotations: can crop sequences explain arable weed seedbank abundance? Weed Research, 51, 422–432.

the level of biological control constitutes a measurable variable which has been the focus of previous experimentation, and accordingly for which databases are available.

The two cases examined were the following:

• the regulation of weed seeds by carabids in wheat crops, assessed in terms of the abundance of granivorous and omnivorous beetles (assuming that an increase in beetle numbers will result in an increase in predation rates for weed seeds);

• the regulation of aphids in wheat, barley, cabbage, and soybean crops, assessed in terms of the difference in aphid growth rates in the presence and absence of natural predators.

In both cases, the methodology was constructed according to the following procedure:

1. A review of the scientific literature was used to identify the major landscape characteristics likely to explain the indicator variable for the level of biological control.

2. Using datasets obtained from prior experimental research, a statistical model was developed to describe the relationships between landscape characteristics and levels of regulation. The model was of the following type:

Variable for the level of regulation = f (landscape characteristics)

3. The model thus developed was then applied to the data from the French LPIS for mainland France, describing agricultural land use at a detailed level. In this way, it was possible to predict potential levels of regulation for each 2 km square cell, and then to make a preliminary assessment of the reliability of the results.

Results

The elements presented below are the result of exploratory work seeking to apply, to the whole of France, models developed using data that are not necessarily representative of the pedoclimatic and agronomic conditions of the entire country (data collected in other countries or data restricted to a specific French administrative Department). It was not possible to further validate these methods during the EFESE-AE project. These very preliminary results are thus intended primarily to illustrate the potential of these methods and to provide an example of the kinds of results they can produce. They should be interpreted with caution, with due recognition of the assumptions underlying the assessment framework.

Table 2-3 summarises the constituent elements for these two methodologies.

Table 2-3. Key elements of the methodologies proposed in EFESE-AE to assess two cases of conservation biological control

Variable to predict (= Indicator of ES level)	Key landscape characteristics identified, based on a review of the literature	Experimental datasets used to develop the model
Abundance of granivorous and omnivorous carabids	- Percentage of land in major field crops within a radius of 1 km* from the center of the field unit - Percentage of land in permanent grassland within a radius of 1 km* from the center of the field unit	Petit *et al.* (2017)[27]: experiment covering 31 fields of winter cereals (barley and wheat) on 13 farms in Côte d'Or
Difference in the growth rate of aphids in the presence and absence of natural predators	- Percentage of land in cultivation within a radius of 1 km* from the center of the field unit	Data compiled by Rush *et al.* (2016)[28]: 15 experiments covering a total of 175 fields of wheat, barley, cabbage, and soybean in 5 countries in Europe and North America

* The spatial range of 1 km was used because previous work has demonstrated its relevance for certain trophic interactions involved in the population dynamics of insect pests and their natural enemies.

Regulation of weed seeds by crop rotations

Figure 2-8 shows the minimum (A), average (B), and maximum (C) abundance of the weed seed bank, as predicted by applying the model developed by Bohan *et al.* (2011) to data for France as a whole. The model design assumes that crop rotations in France have an equivalent effect on the weed seed bank as crop rotations in Great Britain. In addition, it is assumed that varietal differences within a single crop species have no effect not accounted for by the model. Finally, we should note that the applicability of the model is limited to "conventional" cropping systems based on tilling (the prevailing agricultural system type in France).

The highest levels of weed seed abundance predicted by the indicator correspond to major field crop areas in the mid-Garonne basin, in Poitou, Brittany, and Normandy, in the northern part of the Parisian basin, on the Alsatian Plain and in the Rhône Valley. For the vast majority of assessment units (2 x 2 km), the minimum and maximum values estimated by the statistical model correspond to minimum and maximum values

27. Petit S., Trichard A., Biju-Duval L., McLaughlin O.B., Bohan D.A., 2017, Interactions between conservation agricultural practice and landscape composition promote weed seed predation by invertebrates, Agriculture ecosystems and environment, 240 : 45-53.
28. Rusch A., Chaplin-Kramer R., Gardiner M.M., Hawro V., Holland J., Landis D., Thies C., Tscharntke T., Weisser W.W., Winqvist C., Woltz M., Bommarco R., 2016, Agricultural landscape simplification reduces natural pest control: A quantitative synthesis, Agriculture, Ecosystems & Environment, 221 : 198-204. DOI : 10.1016/j.agee.2016.01.039

for the totality of the national dataset. This would seem to suggest that a variety of different weed management objectives, from limiting weed flora as much as possible to regulating weed density within a strategy of biodiversity conservation, can be achieved throughout France by means of crop rotation design.

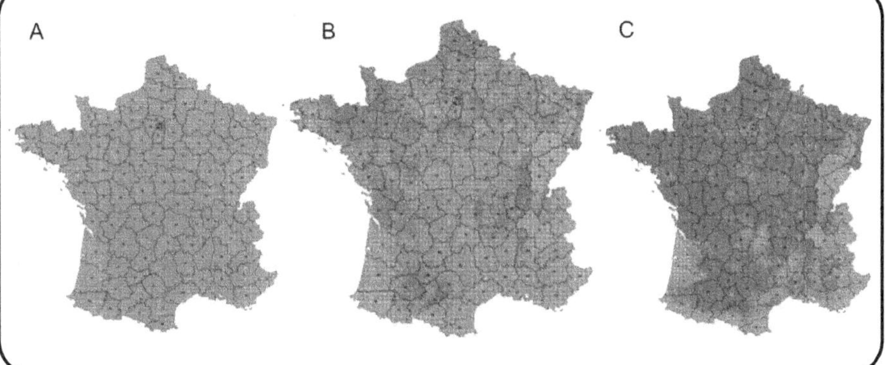

Figure 2-8. Spatial distribution of the estimated minimum (A), average (B), and maximum (C) total soil weed seed abundance (number of seeds per m²) in France

A: Minimum abundance; B: Average abundance; C: Maximum abundance. Number of weed seeds varies from 1 (lower limit of the green range) to more than 10,000 seeds per m² (upper limit of the red range); Resolution: 2 km square cells; Gray pixels: no value (no data available or crop rotation including more than one year in temporary grassland). The minimum and maximum values correspond to minimum and maximum estimated values at the level of the field parcel within each 2 km square cell.

Biological control by natural predators

With regard to the regulation of weed seeds by carabids in wheat crops, the results obtained seem to indicate higher beetle abundance in wheat fields in the mid-Garonne basin, the foothills of the Pyrenees, Poitou-Charentes, Normandy, the Parisian basin (with the exception of Sologne), the Alsatian Plain, and the Rhône Valley. It should be noted, however, that these areas are also those associated with the highest levels of uncertainty. This could be explained by the fact that carabid abundance depends on the presence of both crops and permanent grasslands in the surrounding landscape, but these conditions are not often found together. In situations where only one of these two landscape elements is strongly represented in the landscape matrix, estimated carabid abundance may be high but predictability is uncertain. Furthermore, although omnivorous and granivorous carabids seem to be distributed similarly overall, except in the eastern half of the Massif Central where the indicator predicts more omnivorous carabids than granivorous carabids.

With regard to the regulation of aphids in crops of wheat, barley, cabbage, and soybeans, by design, the results predict the highest levels of regulation in landscapes with a low percentage of cultivated land and a large percentage of forests and/or grassland areas; or in other words, in landscapes considered to be the most "complex" (or heterogeneous). Conversely, and again by design, the lowest anticipated levels are found in the major agricultural production basins, characterised by relatively intensive crop production. However, a high level of uncertainty is associated with geographic areas showing a high level of anticipated regulation (corresponding to the major mountainous areas, with the exception of the Massif Armoricain). Furthermore, these results should be considered in light of the fact that the relative importance of different predator and parasitoid guilds varies significantly according to the pedoclimatic context, particularly between Northern, Central, and Western Europe. Interactions between landscape simplification and levels of regulation may be affected by these differences, and thus verification and calibration according to the different situations encountered in France is needed.

Validation of methods and perspectives for future research

The models previously described need to be verified across a wide range of French pedoclimatic and agronomic contexts. Inra (UMR-BAGAP) has created a database based on a long-term carabid monitoring program on multiple sites across France, which could be used for this purpose with regard to the indicator for the regulation of weed seeds. In addition to this necessary validation of the models, the methodologies presented above should be supplemented by other work to provide a more complete understanding of the ES of conservation biological control. Among other aspects, an assessment of weed seed predation by birds and mammals is needed. The abundance of bird species known to contribute to weed seed consumption could be estimated and mapped using the database created by the French Temporal Monitoring of Common Birds project (*Suivi Temporel des Oiseaux Communs*, or STOC). In addition, other pest/crop pairs could be examined, such as vine moth/winegrapes, pollen beetles/oilseed rape or codling moth/apple and pear trees.

Future verification and supplemental studies thus depend in part on the completion of major campaigns of data collection and data analysis. That being said, a number of innovative biomonitoring methods are currently in development that could complement or eventually even replace the use of traditional field sampling. Next-generation DNA sequencing can be used to quickly determine an inventory of species present in a given location. Applied at the national or even continental level and used in combination with machine learning, these methods may make it possible to assemble data to identify the trophic networks underlying ES of conservation biological control. The effectiveness of this methodology has been demonstrated for fungal and bacterial associations present on the leaves of plants. Such approaches should make it possible to compare networks of ecological interactions across different ecosystems and at a variety of temporal and spatial scales.

Future research is also needed to assess the relative importance of agricultural practices and landscape characteristics in determining the level of conservation biological control. Such research could consider the importance of different cropping systems (organic agriculture, conservation agriculture, etc.) and different landscape types within a given pedoclimatic context. Some recent studies in this direction suggest that landscape effects may be sharply reduced (or even largely eliminated) by the implementation of certain cropping system types. A key limitation for the development of this type of analysis is the availability of detailed data on agricultural practices and landscape characteristics. Data collection procedures for the assessment of the different maps described above should thus also include these elements. Analyses of the effect of these variables on ES expression could lead to further refinement of the models, for instance to account for additional landscape characteristics (e.g., to divide "field crops" and "permanent grassland" into sub-categories based on crop type or the number of years in grass) or cropping system characteristics.

We should note too that current research on the interactions between semi-natural habitats and natural regulation relies on relatively simplistic descriptions of landscapes based on major land use types. More detailed descriptions of these habitats, with an emphasis on their functional aspects (e.g., over-wintering sites for natural predators, alternative food resources), seem essential to the development of a more precise understanding of how levels of natural regulation vary across the landscape.

Eventually, future research on the analysis of biophysical determinants and external factors for ES of conservation biological control should make it possible to define weed- and pest-management strategies allowing for a significant reduction in pesticide use. A better understanding of the effects of different cropping systems (the spatio-temporal configuration of plant cover + external agricultural practices), different landscape configurations, and the interactions between the two is needed to design territorial organisations of agriculture allowing for high levels of ES of biological regulation.

3. The production of agricultural goods: plant and animal products

In the EFESE-AE framework, agricultural goods are understood as a co-production of ecosystem functioning (input ES) and external agricultural practices (farmer-supplied inputs such as fertilisers, soil amendments, irrigation water, energy for soil structuration, and crop protection products). An original methodology for estimating the relative contribution of natural vs. anthropic factors in primary agricultural production, along with the study's preliminary findings, are outlined in the first part of this Chapter.

Estimating the percentage of animal production outputs attributable to agricultural ecosystem services presents additional challenges, both conceptual and methodological. Livestock may be moved from one place to another, and may be fed with plant materials originating from various, sometimes distant, geographic locations. Notwithstanding these challenges, we propose here to estimate the percentage of livestock agriculture outputs attributable to local agricultural ecosystems through an estimation of the percentage of livestock diets (primary plant materials) obtained from within the SAR where they are located, vs. the part of their diet that is imported from elsewhere.

Relative weight of input ecosystem services in the production of plant products

Despite the fact that this question is not unique to agricultural ecosystems, few studies to date have sought to estimate the relative contribution of natural vs. anthropic production factors in the production of agricultural goods. No consensus exists as to the best methodology for partitioning and allocating these different factors. In EFESE-AE, this assessment relies on the dynamic simulations of ecosystem functioning using the STICS model performed to quantify ES relating to the water cycle, the N cycle, and the C cycle (see the previous Chapter). The results of these simulations made it possible to propose a preliminary assessment of the percentage of production outputs attributable to agricultural ecosystem functioning (for six major crop categories), as well as an estimation of the contribution of irrigation and fertilisation practices in meeting crop requirements for water and N (respectively). A major advantage of this type of simulation is that it takes into account daily evolutions in the interactions between external inputs and ecological processes. On the other hand, the model was not able to simulate the effects of crop pests or crop protection practices (see Box 2-1). As a result, only those processes relating to abiotic cycles were represented.

■ Quantifying the relative contribution of input ecosystem services to crop production

The assessment methodology presented below seeks to quantify the percentage of plant production outputs made possible by input ES relating to the supply of N and water to crop plants (referred to hereafter as the input ES "N and water"), as an annual average, assuming a given initial condition for the agricultural ecosystem.

The goal of this process was to establish the initial steps of the theroretical and methodological approach and to determine some preliminary values. Given their preliminary nature, the results should be applied with caution, and with due recognition of the underlying assumptions and methods.

Assessment method

To carry out this assessment, two simulation scenarios were performed, using the STICS model, for the 30 climatic years 1984 to 2013, for the specific PCU defined by the study (see Box 2-1), and based on the same agricultural ecosystem configurations (crop rotations):

- simulation of cropping systems "with inputs": that is, managed with median fertilisation practices for the region; + irrigation of maize to cover 85% of crop requirements; + incorporation of crop residues
- simulation of cropping systems "without inputs": no fertilisation, no irrigation, no incorporation of crop residues. From a theoretical point of view, simulation of these system types – with no external inputs of materials or energy (for soil structuration or the incorporation of crop residues) – is close to so-called "natural agriculture", without human intervention (except for sowing). This simulation makes it possible to estimate the average annual amount of production attributable to the input ES "N and water," assuming a given initial state (levels of organic C and N).

The underlying assumption for this analysis is that comparing annual average yields simulated according to these two scenario types will provide an estimate of the percentage of production attributable to the input ES "N and water", taking into account the properties, characteristics and initial organic condition of soils (soil organic matter levels); climatic conditions over the 30 years of the simulation; and "current" practices with respect to fertilisation, irrigation and crop residue management. The indicator examined below is the average ratio (over the simulation period) of yields without to yields with inputs. It has been calculated for each crop (the average of ratios of crop-year pairs), as well as at the level of the crop rotation (the average of ratios for all crop-year pairs).

Results by crop

As can be seen below, there is considerable variability in the indicator values depending on the crop in question (Figure 3-1). At the national level (all "major field crop" PCUs included in the analysis), sunflower and forage maize show the highest values, suggesting that on average, more than 2/3 of their production is attributable to the input ES "N and water." In 95% of the simulated cases for these two crops, the part of production enabled by the input ES "N and water" is between 50 and 83%. Soft wheat also shows relatively high indicator values, ranging essentially from 47 to 68% (95% of values) with an average of 57%. For grain maize, the indicator shows contrasting values (from 10 to 76%) depending on the geographic area being simulated; the average value is around 41%. Sugar beet and oilseed rape show lower values, averaging 34 and 28%, respectively, and rarely exceeding 40%. For grain maize, the level of ES for water resupply is mostly likely the limiting factor (see below), whereas for oilseed rape and sugar beet, the part of production attributable to the ES "N and water" is more likely to limited by the ES for supply of mineral N.

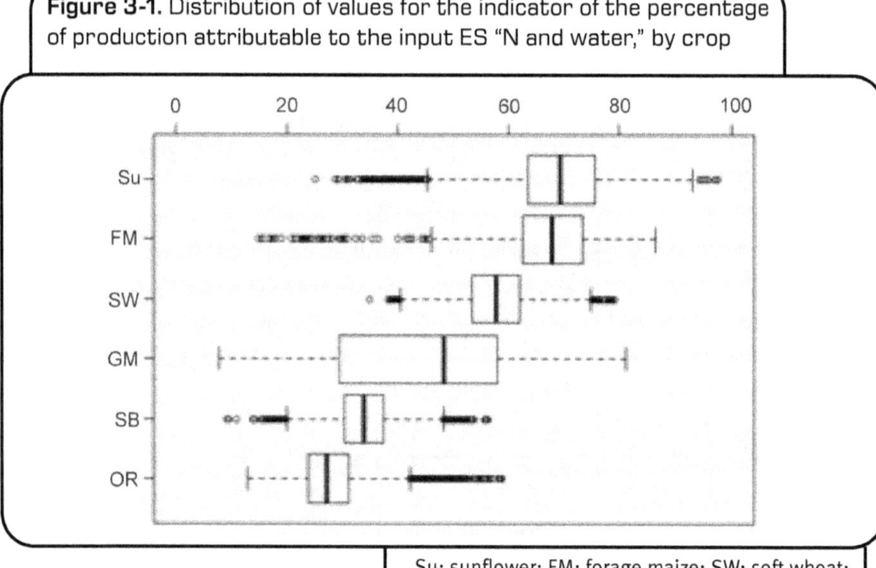

Figure 3-1. Distribution of values for the indicator of the percentage of production attributable to the input ES "N and water," by crop

The following pages provide additional details with respect to two of these crop species: (i) soft wheat, the most widely grown crop in France, and (ii) maize, for which the indicator shows highly variable values depending on the type of crop (grain maize vs. forage maize) and on the pedoclimatic context. The variability of the results was analysed as a function of climate type for each "major field crop" PCU (Figure 3-2).

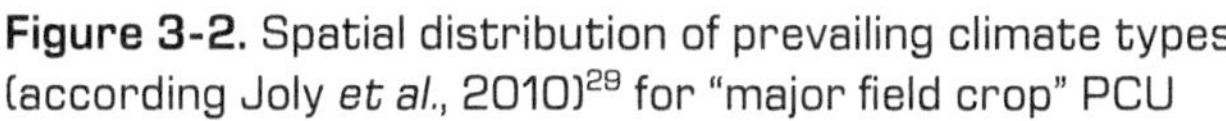

Figure 3-2. Spatial distribution of prevailing climate types (according Joly *et al.*, 2010)[29] for "major field crop" PCU

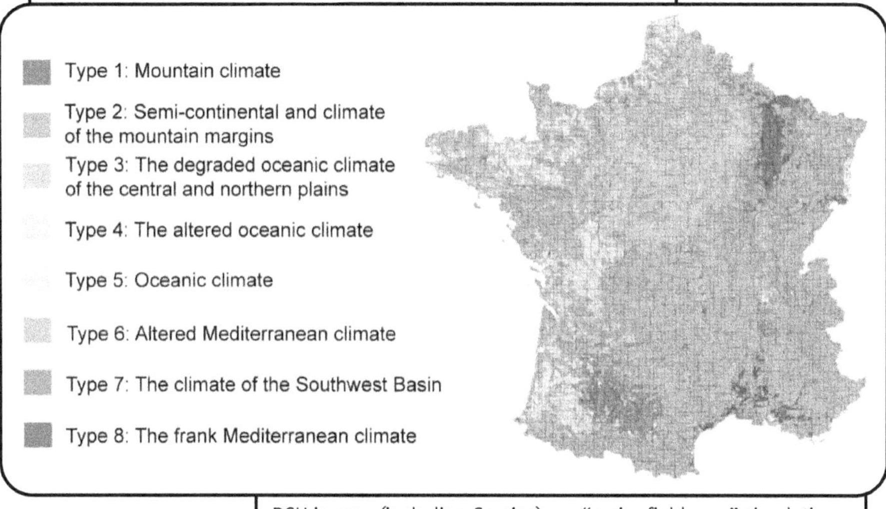

PCU in gray (including Corsica): no "major field crop" simulations.

Example of soft wheat

Figure 3-3 shows the spatial distribution of values for soft wheat production made possible by the input ES "N and water." Analysis of the results by climate type reveals that the highest average levels for the part of production attributable to the input ES "N and water" are observed in climate types 6 and 7 (60%) and in climate type 8 (65%). Variability is similar across the different climates, although slightly less in climate type 8. The findings observed in the Mediterranean climate zones (6 and 8) and in the Southwest (7) may reflect lower yield potentials in these areas compared to the major French cereal growing region. In other words, a given level of ES "N and water" will meet a higher percentage of crop needs in situations where yield potentials (and thus crop N and water requirements) are low.

The highest average ratios are observed on coarse and on fine-grained soils. Here again, this effect may be attributable to lower yield potentials in these contexts.

Although crop rotation characteristics (rotation length) do not appear to differentiate the indicator values, the part of production attributable to the input ES "N and water" tends to be higher where cover crop use is low (the median value is around 55-60% when the cover crop rate is low vs. around 50% for the highest rates). This cause of

---

29. Joly D, Brossard T, Cardot H, Cavailhes J, Hilal M, Wavresky P. (2010). Les types de climats en France, une construction spatiale. *Cybergeo: European Journal of Geography.* https://doi.org/10.4000/cybergeo.23155

this effect and its behavior with respect to the characteristics of different cropping systems need to be examined in more detail.

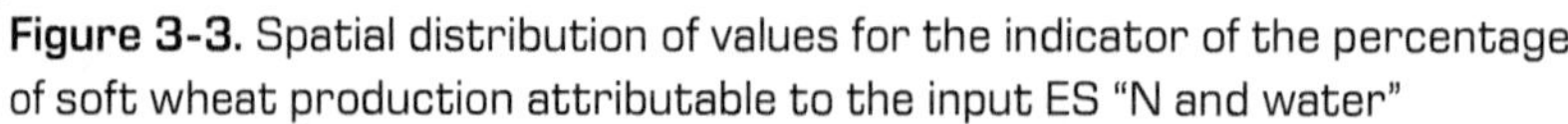

Figure 3-3. Spatial distribution of values for the indicator of the percentage of soft wheat production attributable to the input ES "N and water"

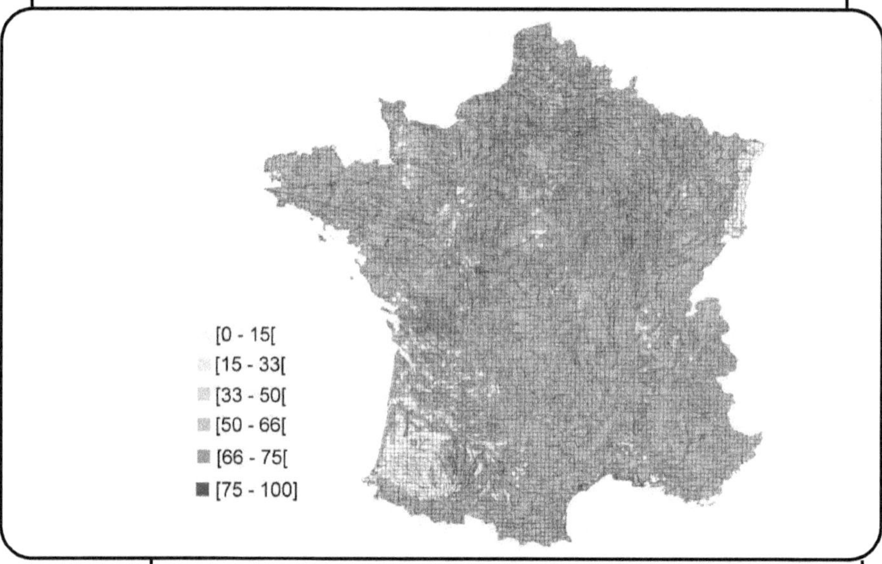

Spatial resolution: PCU; PCU in gray (including Corsica): no "major field crop" simulations; PCU in white: soft wheat not simulated in the rotations.

Finally, the fact that wheat is simulated without irrigation in the current systems favors a higher result (that is, a higher estimate of the percentage of production attributable to the ES "N and water"). Wheat crop irrigation, which is becoming increasingly common in France, especially for hard wheat, can produce significant yield increases in regions marked by a high water deficit. As calculated here (ratio of simulated yields without and with inputs), it is reasonable to assume that the part of production attributable to the input ES "N and water" would be lower if the reference scenario (current systems with inputs) included irrigation for wheat.

Example of maize

Figure 3-4 shows the spatial distribution of values for the percentage of maize production (grain and forage maize) made possible by the input ES "N and water."

For grain maize, the average level of the part of production attributable to the ES "N and water" is highly variable across climate types: approximately 55-60% in climate types 1, 3, and 5 (mountain, degraded oceanic and oceanic), roughly 45% in climate 2 (semi-continental), a bit below 30% in climate 6 (altered Mediterranean) and around 20% in climates 7 and 8 (Southwest and frank Mediterranean). Variability within climate types is higher where ES-supported production levels (in %) are lower. Crop stress associated with low average moisture availability in the summer and/or high year-to-year variability in moisture conditions appear to be the principal factors underlying this variation. These observations should be considered in light of the fact that most grain maize is grown in the hottest parts of France, which are frequently also the areas with the lowest rates of summer rainfall and thus areas where irrigation is widely used (the greater Southwest, central France and Alsace). No major soil effect was identified.

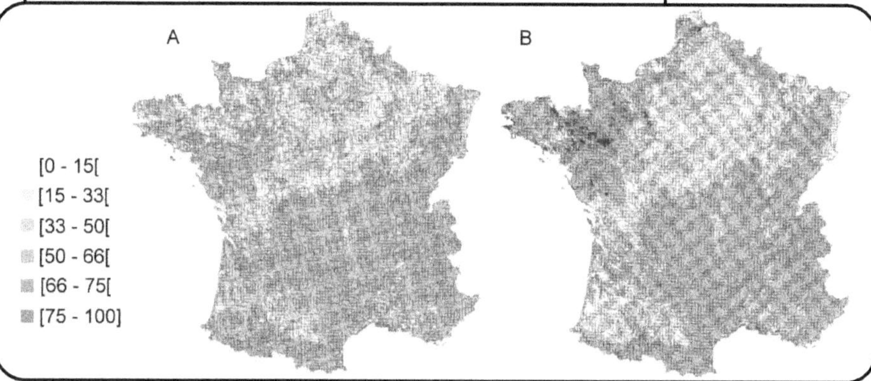

Figure 3-4. Spatial distribution of values for the indicator of the percentage of grain maize (A) and forage maize (B) production attributable to the input ES "N and water"

A: Grain maize; B: Forage maize. Spatial resolution: PCU; PCU in gray (including Corsica): no "major field crop" simulations; PCU in white: grain maize not simulated in the rotations.

Increased use of straw cereals and/or cover crops within crop rotations was also associated with higher values. Here again, however, these conclusions must be considered in light of the distribution of crop rotations and cover crop use across different climatic zones (possible confusion of crop rotation effects and climate effects given the distribution of the former across the latter).

For forage maize, in contrast to grain maize, the part of production attributable to the ES "N and water" varies little, and is generally 60 to 70%, with levels slightly higher in climates 4, 5, and 6 (altered oceanic, oceanic and altered Mediterranean), which are somewhat better suited to the needs of forage maize as a crop. These results are directly linked to the geographic distribution of forage maize, which is highly concentrated in the oceanic climate zones. Results may also reflect the fact that the water requirements of forage maize are slightly lower than those for grain maize, given its shorter growth cycle. No other major effect was identified for this crop.

Results at the scale of the crop rotation

At the scale of the cropping system, the part of production attributable to the input ES "N and water" appears to be approximately 50% on average, with 95% of values varying from 29% to 71%.

Cropping systems associated with higher values are located in the Garonne basin, in the southern half of the Rhône basin, and in Brittany (Figure 3-5). The Landes region and the Alsatian plain are associated with lower values, mainly linked to the prevalence of continuous maize. As with wheat and maize, the variability of the results was analysed as a function of climate type (Figure 3-6).

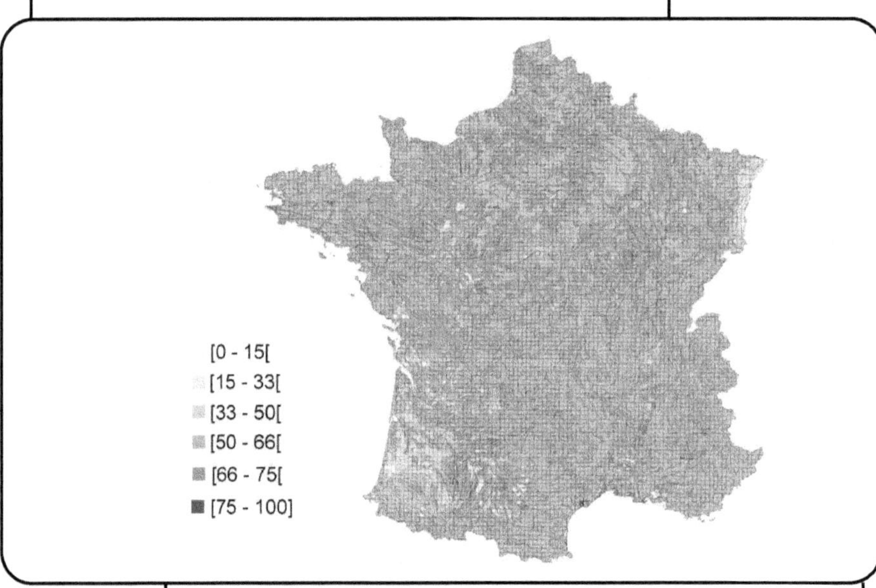

Figure 3-5. Spatial distribution of values for the indicator of the percentage of crop production attributable to the input ES "N and water" at the scale of the rotation

Spatial resolution: PCU; PCU in gray (including Corsica): no "major field crop" simulations; PCU in white: excluded from the analysis.

Climate types 1, 2, and 3, corresponding primarily to northern and northeastern France, are associated with values of approximately the same level: for the majority of PCU located in these climate areas, the part of crop production attributable to the input ES "N and water" varies between about 30% and 70%, with the median value around 50%. In these climates, simulated crops are primarily soft wheat, oilseed rape, grain maize (in the southern part of the zone and in Alsace, mainly as continuous maize), and sugar beet (in the Hauts-de-France region and in Champagne).

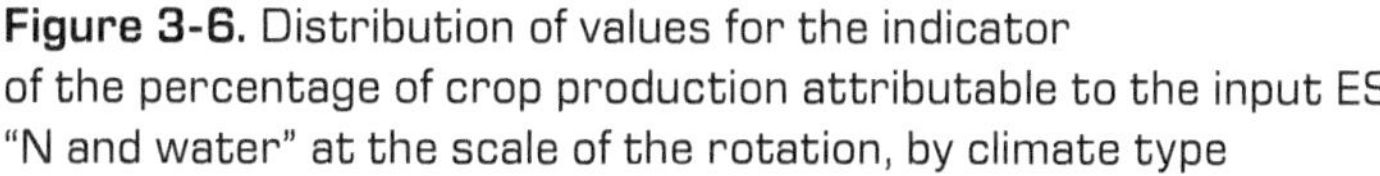

Figure 3-6. Distribution of values for the indicator
of the percentage of crop production attributable to the input ES
"N and water" at the scale of the rotation, by climate type

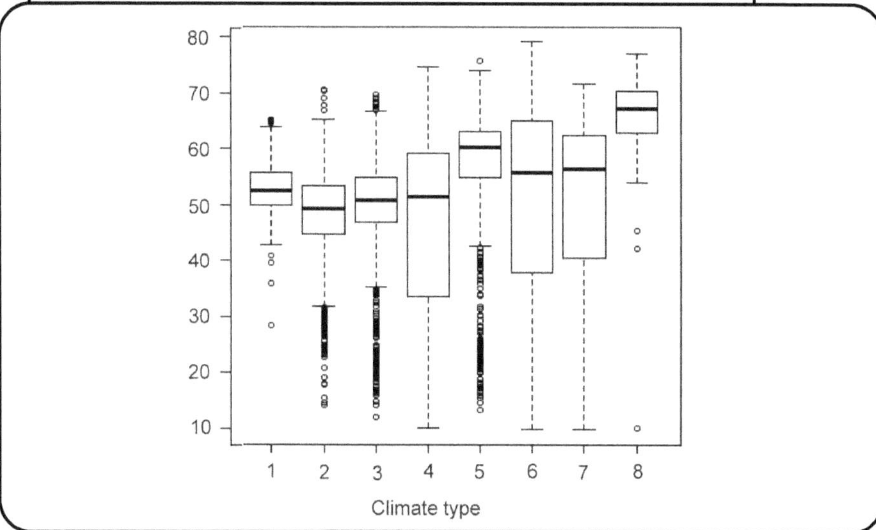

Climate types 4, 6, and 7 – "transition" climates lying between central France and the coastal areas – are associated with a high variability of indicator values, covering almost the full range found in France as a whole. The wide diversity of crops, cropping sequences, and soil types found in PCU of climate type 4 (altered oceanic), type 6 (altered Mediterranean), and type 7 (Southwest) explain this variability. Thus, for example, in the greater mid-Garonne basin (climate type 7), simulated rotations included wheat-(wheat)-sunflower and wheat-sunflower-wheat-oilseed rape on the clay-limestone hillsides, and continuous (grain) maize on the primarily loamy soils along the Garonne River. This wide range of crop-soil-climate combinations, including crops with different seasonal peaks in terms of water and N requirements, produces a variability in ecosystems' capacity to supply N and water to crops across these different climates.

Values associated with PCU in climate type 5 are more concentrated in the 55-65% range, rarely falling below 40% and never exceeding 75%. These PCU, found throughout the Atlantic coastal area, likewise include a wide range of cropping systems. The preponderance of forage maize cultivation, and to a lesser extent grain maize, helps to explain these high overall values.

Finally, PCU located in climate type 8 are simulated almost entirely as continuous wheat. These show the highest indicator values, ranging for the most part from 53-77%. As explained above, these high values for the part of production attributable to the input ES "N and water" may be explained by the relatively low yield potential of wheat in these geographic areas.

In general, due to the nature of the calculation, the results reviewed here are directly linked both to the geographic distribution of the various crops (see above) and to their temporal distribution within crop rotations. The nature of this relationship remains to be analysed in more detail.

▌ Estimating the relative contribution of input services vs. agricultural practices in meeting crop requirements

Relative contribution of the input ES "water" vs. irrigation in supplying water requirements for maize

To estimate the contribution of the input ES "water" vs. irrigation in meeting crop water requirements (in this case, maize), the ratio between the annual quantity of water supplied by the ecosystem (transpiration - irrigation) and the annual quantity of water used by the crop (water transpired by the crop) was calculated for each PCU containing irrigated maize.

On average across all PCU considered in the analysis, input ES "water" contributes 37% of the water requirements for maize. Contexts in which this ES contributes the most (from 50-80%) to maize crop water requirements are found in PCU in the foothills of the Atlantic Pyrenees and on the Alsatian plain (Figure 3-7). Situations in which it contributes the least (less than 30% on average) are found in the Landes, in Poitou-Charentes and in the Rhône basin. Irrigated maize crop grown along the length of the Garonne, along the Gascogne and along the Adour occupy an intermediate position with respect to ES meeting crop water requirements. Here again one can observe a direct link with the average summer water deficit of these different climatic zones: in locations where dry summers are more pronounced, the input ES "water" is less able to meet the water requirements of maize, and thus irrigation becomes more important.

PCU characterised by sandy soils or very clayey soils, with a low maximum available water reserve, and located in more arid climatic zones, show an important role for irrigation in meeting the water requirements of maize (75% on average). Soils in these areas are only able to retain small amounts of water during the rainy season for use by crops through their main period of growth, during the dry part of the year.

Finally, in approximately 20% of PCU, irrigation supplies nearly all of crop water requirements on average (PCU shown in red in Figure 3-7). The biophysical determinants and external factors involved in these situations have yet to be analysed.

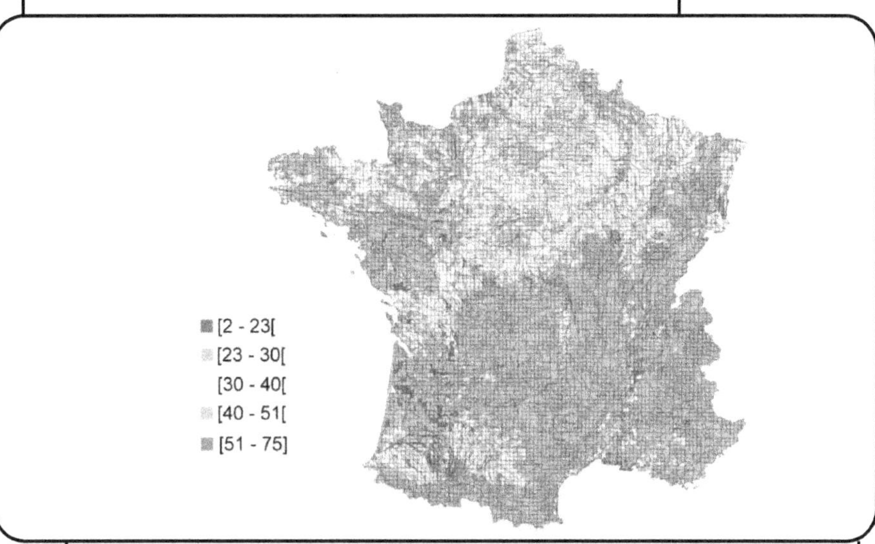

Figure 3-7. Spatial distribution of the percentage of water requirements of maize attributable to the input ES "water"

Spatial resolution: PCU; The simulations were only carried out for PCUs in which maize is typically irrigated; PCU in gray: no "major field crop" simulations.

Contribution of the input ES "N" vs. fertilisation in meeting crop N requirements

To estimate the relative contribution of the ES "N" vs. fertilisation in meeting crop N requirements, the ratio was taken of the quantity of N supplied by the ecosystem and through external inputs to the quantity of N present in the commercial crop at harvest. Three variants for the numerator of this ratio were considered:

• Ratio 1: the quantity of mineral N supplied by the ecosystem during the period of crop growth (from planting to harvest), without accounting for the amount of mineral N in the soil at the time of sowing. This provides an analysis of the relative importance of mineral N supplied by the ecosystem in meeting crop needs during the growing season.

• Ratio 2: the quantity of mineral N supplied by the ecosystem during the period of crop growth including the quantity of mineral N in the soil at the time of sowing. This provides an analysis of the relative importance of mineral N supplied by the ecosystem during crop growth and prior to crop growth in meeting crop requirements. The effect of fertilisers applied during the preceding crop cycle on the stock of mineral N in the soil at the time of sowing is thus included in this analysis.

• Ratio 3: the quantity of mineral N supplied by the ecosystem during the period of crop growth including the amount of mineral N in the soil at the time of sowing, plus the addition of external N inputs (fertilisation). This provides an analysis of whether the total of all sources of mineral N (as supplied by the ecosystem prior to and during the crop cycle + as supplied by fertilisation) is greater than, equal to, or less than crop requirements.

Calculation of ratio 1 shows that during the crop cycle, the ecosystem supplies, on average, 40-50% of crop N requirements. Soil contributions are lower where soil clay content is lower. Areas in central France and the Southeast show the lowest ratios, while the major areas of grain maize and forage maize production – the Southwest, Alsace and Brittany – show the highest ratios (Figure 3-8).

Figure 3-8. Spatial distribution of the percentage of crop N requirements attributable to the input ES "N"

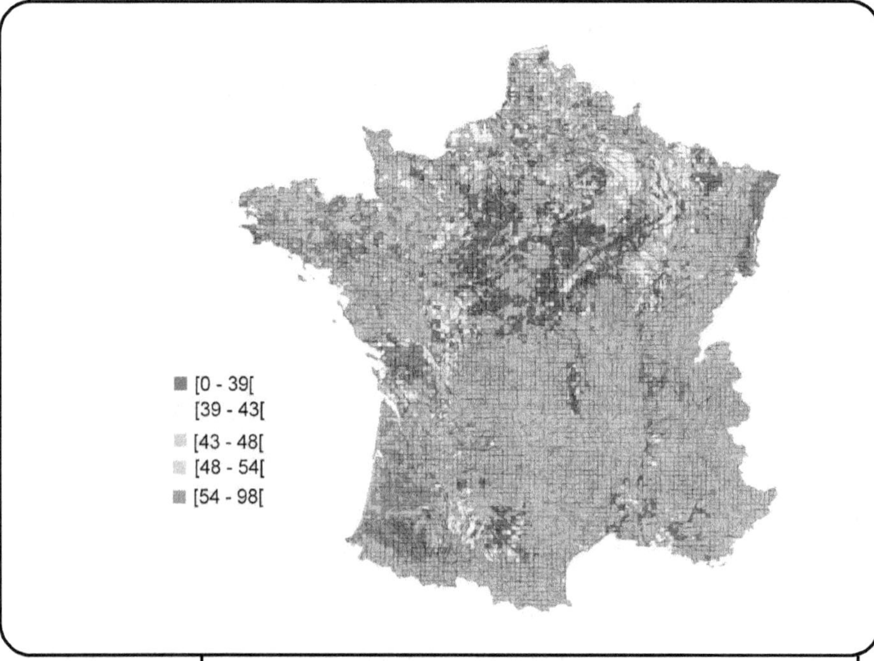

Spatial resolution: PCU; PCU in gray: no "major field crop" simulations; PCU in white: excluded from the analysis.

If one takes into account the mineral N present in the soil at the time of sowing (ratio 2), the contribution to crop N requirements is around 75% on average across all soil types, with a slightly lower contribution observed on very clayey soils (soils with a clay content of greater than 60%, which are rarely used for major field crop production and are thus only rarely present in the simulations). For some simulations, the ratio is greater than 1, suggesting that there are situations (for some crop rotations, in some soil types and in some climate types) in which crop N requirements are, on average, met by N available in the soil at the time of sowing + as supplied by the ecosystem during the crop cycle. The precise characteristics of these situations remain to be identified.

Finally, if one takes into account all external N inputs (the remainder of what is supplied in year n-1 in addition to what is supplied in year n) the average ratio is well above 1, and no values are found below 1. Across all simulation variables (crop rotation, soil type, climate type) the average amount of mineral N available for plant growth is in excess of crop requirements. Given that the N inputs used by the simulation tool are the median of N inputs for the administrative region as reported in the Agricultural Practices Survey for 2011, these results suggest that on average, at the level of the cropping system, a reduction in fertilisation applications (mineral and/or organic) could be made. This finding is in line with the fact that farmers' fertilisation practices are in most cases based on the "balance method" (either partial or complete); which is to say, based on meeting crop needs according to yield objectives that are not always achieved (e.g., 2 years out of 5). Such an analysis needs to be qualified, however, with respect to two major limitations of the simulation tool. First, sowing and fertilisation dates are assumed to be uniform for all simulation units within an administrative region, and thus are not tailored to specific soil types or annual climate regimes. It is thus possible that, in some years, sowing dates and N inputs are poorly adapted to weather conditions and thus that crop development is not represented accurately. In this situation an underestimate of crop development leads to an underestimate of the amount of N used by the crop. The second issue is that because the level of crop fertilisation is held constant for all PCU within a region, it is possible that these levels are too high for simulation units with low yield potential (i.e., marginal crop areas). The frequency of this latter phenomenon remains to be determined, but is *a priori* low within regions.

A more detailed analysis of cropping systems with ratios greater than 1 is needed to further assess the potential for reducing nitrogenous fertilisation rates. A temporal analysis of changes in N flows over the course of the cropping season would also be useful. It is possible that, on an annual basis, N availability in the soil is asynchronous with periods of crop need and thus that N is being lost, for instance by leaching beyond the root zone.

Quantifying the percentage of livestock production relying on locally produced plant resources

Livestock on pasture are considered in EFESE-AE as biotic components of the agricultural ecosystem. By consuming plant resources and producing manures, livestock animals present in the ecosystem (both ruminants and non-ruminants) play a major role in flows of material and energy within the ecosystem. In addition to their role as biophysical determinants in the supply of some ES, livestock are involved in the production of secondary-level agricultural goods by using some quantity of primary (plant) production.

Livestock production makes use of a variety of plant resources depending on the animal species and the type of production system. Animals' feed conversion efficiency can also be highly variable, depending on plant resources (cereals, oilseed/protein crops, pasture, hay, straw, oilseed cake, pulps, etc.), animal species and breeds (ruminants vs. non-ruminants), and production systems (type of infrastructure, herd management practices, etc.). While it is possible to distinguish various categories of livestock production based on the type of land used for feed production or available on-farm (see Box 3-1), considerable diversity exists within each category, notably in terms of the nature of the relationship between livestock and available land resources, including the percentage of animal feedstuffs produced on-farm and the fate and disposition of animal manures.

Box 3-1. Livestock production categories based on the type of on-farm land area dedicated to animal feeding

Livestock production in buildings, with little or no associated agricultural land area

This includes the majority of non-ruminant animal production (pigs, poultry) in France, and is primarily located in the western part of the country. Animal feeding is based on rations purchased from animal feed suppliers, and land areas for the spreading of manures are more or less distant from the farm. Interactions between animals and land areas are thus mostly indirect.

Mixed crop-livestock production combining livestock with cultivated land

This includes farm operations where the animals (non-ruminants, dairy animals, ruminant animals for meat production) are fed for the most part using crop materials produced on-farm. In this type of production, interactions between animals and land areas may be more or less direct (pasture; animals fed in buildings using plant biomass harvested on-farm; animal manures stored and then spread on farm fields).

Grass-based livestock production based on intensively managed pastures (cultivated grasslands, fertilised permanent grasslands)

This category consists primarily of ruminant livestock production in lowland areas. During the grazing season (from a few months to the whole year), interactions between animals and land are direct. Outside of the grazing season, and sometimes as a complement to grazing, animals are fed with forages produced on-farm and with feeds produced off-farm (concentrates, straw, hay, silage).

Grass-based livestock production based on non-intensively managed grasslands

This includes ruminant production in mountainous areas and pastoral systems in dry parts of the peri-Mediterranean region. These are often environmentally sensitive landscapes and are frequently subject to multiple types of land use (e.g., agriculture, hunting, tourism). Low land productivity means that up to several hundred hectares can be needed for a single farm. On these farms, animals interact directly with plant production. Difficult environmental conditions may impose several months of indoor feeding and thus an indirect relationship between animals and land for part of the year.

A full assessment of the contribution of agricultural ecosystem functioning to the production of animal goods would require developing an estimate of the percentage of primary plant material (PPM) production attributable to regulating ES vs. anthropic inputs for each PPM stream entering into the composition of animal feedstuffs (including PPM imported from other countries). The preceding section proposed an initial assessment of the part of plant production attributable to input ES "N and water" for some major crops. Following the same approach for animal goods poses additional methodological challenges, since animal production can potentially depend on plant production at multiple geographic levels: materials produced on farm, materials produced in the same geographic region as the farm, and materials produced elsewhere (potentially from another continent). It was not possible here to undertake a full assessment of these multiple levels.

Instead, recognising the environmental and economic importance of the link between livestock production and cropland areas, the study undertook a preliminary estimate of the part of animal production relying on locally produced plant materials. The goal was to assess the capacity of agricultural ecosystems situated within the geographic influence of a given territory to satisfy the food requirements of the livestock animals present within that territory.

Assessment method

The contribution of local plant production to the production of animal goods was estimated thanks to a calculation of the ratio between the supply of PPM produced within a territory and the consumption of PPM by animals present in that territory (Figure 3-9). The resulting coefficient corresponds to the percentage of feed consumed by livestock within a territory that could be met by the plant resources produced within that territory. This indicator was calculated at the level of the SAR. All types of livestock animals were considered for their full life cycle (not just the grazing period for ruminants). Five animal species – cows, sheep, goats, pigs, and chickens – and four types of animal goods (milk, meat, animals,[30] eggs) were included in the analysis.

---

30. An animal as an agricultural good is an animal sold live to another farmer to be fed or kept for reproductive purposes.

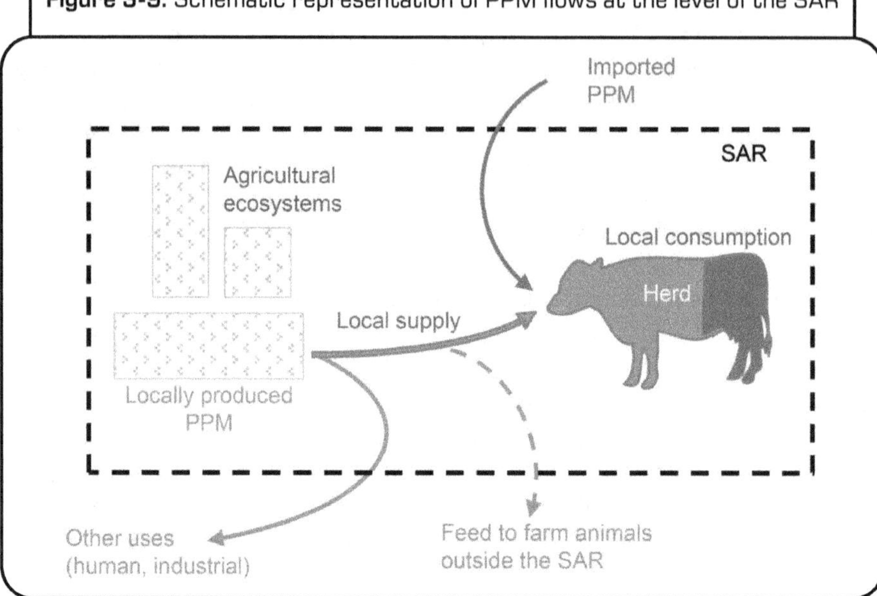

Figure 3-9. Schematic representation of PPM flows at the level of the SAR

Supply: The quantity of PPM produced within the SAR and destined for animal consumption. Different types of PPM contribute to the composition of animal feedstuffs: crop outputs consumed in the form of concentrates; forages; byproducts from food processing or industrial sources; plant goods from crop production.
NB: The potential quantity of PPM produced within the SAR but consumed by animals outside the SAR (green dotted arrow) was not estimated.
All PPM produced for animal consumption within the SAR were assumed to be intended for the feeding of animals within the SAR.
Consumption: Total quantity of PPM needed to feed livestock in the SAR.

Quantification of the local supply of PPM

An estimate of the total production of plant goods by agricultural ecosystems in France was made by combining data on crop yields reported by the Annual Agricultural Statistics, collected for each administrative Department, with data on land area planted to different crops as reported in the Agricultural Census of 2010 (at the level of the SAR). The supply balance established at the national level by the French Ministry for Agriculture, in conjunction with data from FranceAgriMer, make it possible to distribute national production of plant goods to different purposes: animal feedstuffs, human consumption, industrial uses, processing, seed, and waste/loss. The same distribution was applied to all SAR. Since variations in the allocation of plant goods from one SAR to another could not be accounted for, the results for individual SAR should be considered with caution. It is likely that within individual SAR, the share of plant production intended for animal feed is significantly higher or, on the contrary, lower than the national average.

Quantification of PPM consumed by local farm animals

No single database allows for the calculation of total food intake across the five types of farm animals considered here. For ruminants, the *Institut de l'élevage* (Livestock Production Institute) estimated total PPM consumption as an element within its *Autosysel*[31] program, drawing on data from an initiative known as *Inosys-réseaux d'élevage 2008*. Total PPM consumption for non-ruminants was estimated using the Céréopa[32] database.

The ratio of local supply to local consumption of PPM was estimated in terms of i) dry matter and ii) nitrogenous matter. Two variants of the indicator for the local ecosystem's contribution to the production of animal goods were thus calculated; the smaller of these two values was then retained, based on the logic that this would represent the limiting factor for local animal production. For each SAR, the coefficient was then multiplied by the total quantity of animal goods produced within the SAR (calculated from Annual Agricultural Statistics data), to arrive at an estimate of animal production "based on internal SAR resources."

Results

Only those SAR oriented toward livestock production – a total of 571 SAR – were retained for analysis.

The capacity of SAR to supply local animal feed consumption

Figure 3-10 shows the contribution of local plant production to the production of animal goods at the level of the SAR. On average, across the 571 SAR considered in the analysis, locally produced plant resources cover 86% of livestock feed and forage requirements. The total production of animal goods using resources from the agricultural ecosystem itself is thus estimated to be 1.1 million tons of protein. Examination of these results at the individual SAR level reveals a significant heterogeneity of indicator values, varying from 0.25 to 1 (25% to 100%).

Seventy-one percent of SAR have a capacity superior to 0.78. These SAR are found in the grassland plain regions dominated by dairy production (Lower Normandy, Lorraine); in the mixed crop and livestock areas of the Aquitaine Basin and to the east, south, and west of the Parisian basin, where livestock production is in sharp decline; and in some of the grassland areas at the northern edge of the Massif Central, where cow-calf operations are common.

---

31. https://idele.fr/autosysel
32. *Centre d'études et de recherche sur l'économie et l'organisation des productions animales* (Center for the Study of the Economics and Organisation of Animal Productions).

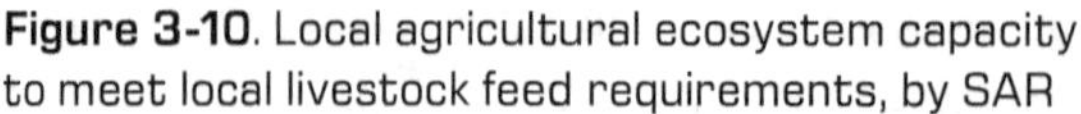

Figure 3-10. Local agricultural ecosystem capacity to meet local livestock feed requirements, by SAR

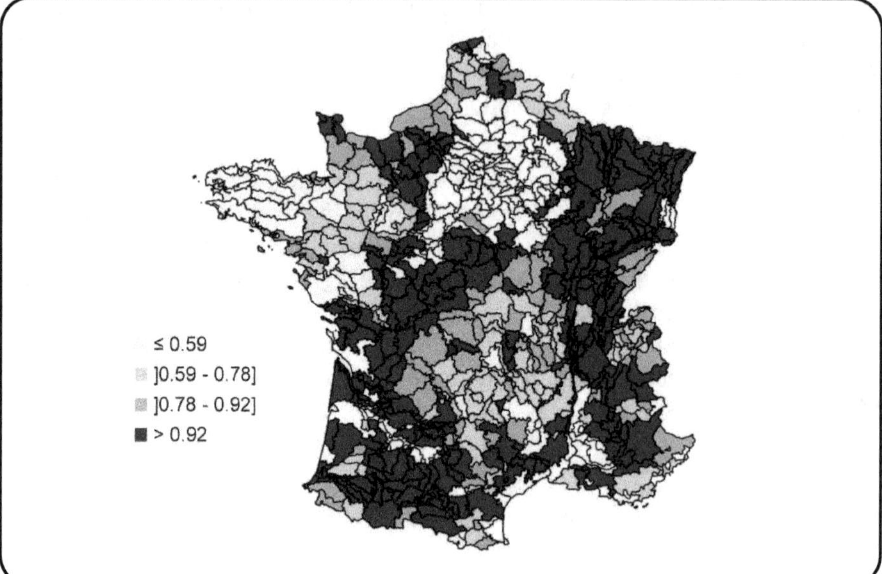

Classification of values in four classes was made using the Jenks algorithm (maximising inter-class variance and minimising intra-class variance). White areas (including Corsica): SAR excluded from the analysis. Example: A capacity of 0.75 means that plant resources produced within the SAR satisfy ¾ of the food consumption requirements of all farm animals present within the SAR.

The remaining SAR have livestock feeding capacities between 0.25 and 0.78. These areas correspond primarily to France's most intensive livestock production zones: i) the Great West (made up of SAR in Brittany, the Loire Valley, and the northwestern part of Poitou), which includes both ruminant and non-ruminant livestock production and cropland dedicated to the production of animal feed; ii) the Massif Central (notably Auvergne), which includes both dairy and meat production, with significant area in permanent mountain grassland. Other SAR with these values are found elsewhere in France, including in the Nord-Pas de Calais region, in the mountainous dairy regions (the Vosges, Franche-Comté, the Alps), and in the pastoral Mediterranean zone.

Herd movements related to the use of pastoral areas are not accounted for in this assessment. SAR situated in areas with transhumant livestock production (the Var, the Crau Plain, the Pyrenees) show a low capacity to meet local animal consumption requirements since they in fact rely on plant production from other SAR during 3-4 months of the year. Around the Mediterranean, not all pastoral land resources used for grazing are subject to CAP declarations, and thus the vegetative output of these areas is often underestimated. In these areas only grassland production is accounted for, whereas livestock also consume the leaves of woody vegetation, sometimes in significant amounts.

Typology of SAR production and capacity profiles

A multi-variant analysis combining a principal component analysis and an ascending hierarchical classification enabled the identification of four groups of SAR based on their capacity to supply local livestock consumption requirements and their total level of animal output (Figure 3-11).

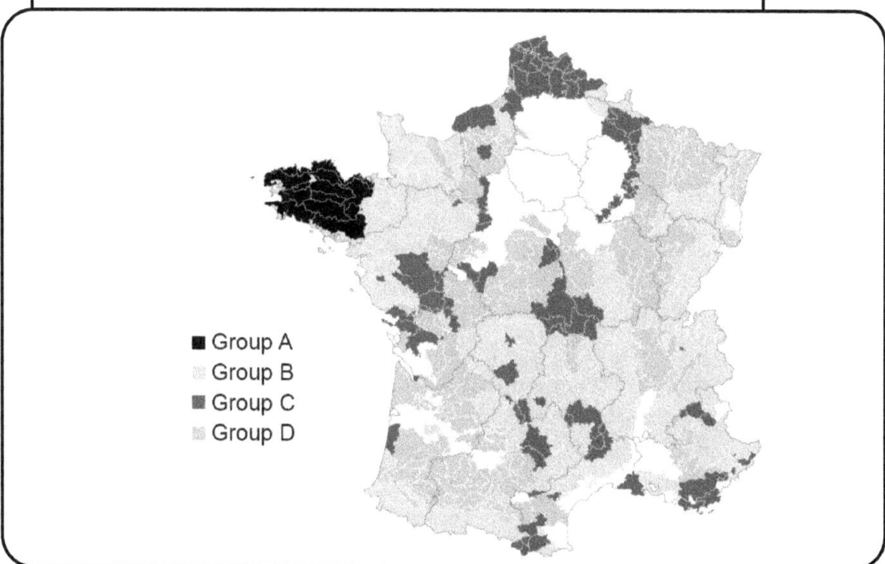

Figure 3-11. Typology of SAR according to their capacity to supply the food consumption requirements of the livestock present and level of production of associated animal goods

Table 3-1 shows the average characteristics of these four SAR groups. An SAR's capacity to supply livestock consumption requirements tends to decrease with livestock densities decrease. Beyond 1.5 LU_{TD}[33] per hectare of agricultural land, that capacity falls sharply due to an imbalance between the number of animals and the available agricultural land area to supply food resources. For all SAR groups, moreover, it is the capacity to supply concentrated protein for animal consumption that is most limiting.

---

33.   LU (livestock unit) "total diet." This unit of measure makes it possible to compare animals based on their total food consumption (bulk materials and/or concentrates), even when they eat different types of feed. One LU_{TD} is defined as a 600-kg dairy cow consuming 3,000 forage units (FU) per year and producing 3,000 kg of milk. An dairy cow thus represents 1.45 LU_{TD}.

Table 3-1. Average features of the four SAR groups, defined in terms of production levels and capacity to meet the food requirements of livestock present within the SAR

Group	Number of SAR	Livestock density (LU_{TD}/ha agricultural land)	Percentage ruminants (% agricultural land)	Level of production (kg protein/ ha agricultural land)	Capacity to meet animal consumption requirements*:		
					in forages	in concentrates	total capacity
A	12	3.6	30%	300	1.00	0.08	0.35
B	250	1.1	86%	70	1.00	0.20	0.81
C	97	0.9	81%	55	0.84	0.60	0.77
D	212	0.6	79%	37	1.00	0.91	0.98
Overall	571	0.9	81%	60.5			0.86

* ratios calculated in terms of total nitrogenous matter.

Results were analysed as a function of indicators of agricultural land composition, thus providing information as to the orientation of livestock feeding systems: land area in cereals and oilseed/protein crops (ACOP), land area in primary forages (APF), land area in forage maize, and land area in permanent grassland (APG)

Group A – A very high production of animal outputs, but dependent on plant material from outside the local area

The 12 SAR that make up this group are all located in Brittany, an area of intensive non-ruminant animal production and intensive dairying. The level of output of livestock goods is high (on average, 114 kg per hectare of agricultural land of proteins from ruminant livestock and 190 kg per hectare of agricultural land of proteins from non-ruminant livestock), and is based on animal densities far above than the overall average, largely due to the numbers of non-ruminant livestock. Group A is thus significantly different from the other groups based on these measures.

The high forage capacity of these SAR is based largely on maize, which is about three times more present within the APF than the overall average. Total capacity is nevertheless well below that of the other groups due to the extreme weakness in the capacity to supply concentrated proteins (8% on average), which are essential as a feed complement to forage maize. The availability of ACOP land per animal is low, again illustrating the extremely high livestock densities and thus the livestock/crop disequilibrium that characterises this SAR group.

Group B – A high level of production of animal outputs made possible by plant agriculture within the territory, primarily forages

This group includes areas of grass-based livestock production in lowland and medium-elevation areas (mostly without transhumance). SAR are spread across four regions. The first includes most of Lower Normandy and the Loire Valley (a dairying area). The second includes most of the Massif Central (Limousin + Auvergne) and its northeastern and southwestern peripheries (cow-calf production and mixed dairy and meat systems). The third, also a dairying area, includes the Eastern Plain and the wet mountain areas (the Vosges, Franche-Comté). The fourth includes the high mountains of the Alps and Pyrenees. The average level of animal production is one-quarter that of Group A and is associated with animal densities slightly higher than the overall average, consisting primarily of ruminants. The very high forage capacity is based on extensive grassland areas, made up almost entirely of APG. Animal density is not the limiting factor for capacity. Although the high level of animal production is based on forages, it also implies a use of protein concentrates in excess of local supplies (as reflected in the ACOP).

Group C – A modest level of production of animal outputs based largely on crop production within the territory

In this group the level of production of animal outputs averages 55 kg of protein per hectare of agricultural land, three-quarters of which comes from ruminants. This group is highly heterogeneous both in terms of the capacity of plant resources to supply livestock feed requirements (varying from 0.25 to 1) and in terms of agricultural land composition. It could thus be useful to refine the typology within this group. Contrary to the other groups, overall capacity is also limited by the capacity to supply livestock forage requirements. This limitation appears to be linked to a low availability of APF relative to the livestock population.

Group D – A low level of production of animal goods based entirely on local plant production

The SAR in this group, located in the area around the Parisian basin and the greater Aquitaine basin, where livestock production has declined sharply over the past 30 years, appear to be almost self-sufficient in plant resources. This result is explained by the low number of animals and thus low level of production. Availability of ACOP is the highest of the four groups, with an average of 1 ha of ACOP per LU_{TD}.

Perspectives for improvement

The results presented above should be interpreted as trends, given the approximations that had to be made in assembling the different data elements included in the indicator. These approximations are summarised below.

First, for each PPM, a coefficient of availability for animal feeding was established at the national level, based on surveys conducted by Agreste and FranceAgriMer. These "average" coefficients, applied indifferently across all SAR, may mask strong regional disparities with regard to the destination of different PPM, thus leading to an overestimation or an underestimation of the local PPM supply. In future work it would be useful to estimate the spatial variability of these coefficients.

Second, no harmonised databases exist for estimating livestock feed consumption across livestock categories. As a result, two different strategies were employed, one for ruminants and one for non-ruminants. The estimate of the PPM requirements for ruminants, in particular, was based on descriptions of average feed rations in use for different types of livestock production systems as reported within the livestock production network of the French Livestock Institute (IDELE). An important future step would thus be to evaluate the representativeness of this network in terms of feeding practices.

Third, the vegetative resources exploited by pastoral production (in the Mediterranean zone and on the plateaus and hills of the Southwest) had a tendency to be underestimated in this initial analysis. In reality, this type of livestock production makes use of brushy and wooded land areas that were not captured in the vegetative resources calculation. In addition, for the sake of simplicity, herd movements were not considered in the analysis, whereas animals belonging to SAR in areas of livestock transhumance may also make use of plant resources beyond the SAR for 3 to 4 months of the year.

Finally and more broadly speaking, the results obtained in this preliminary evaluation highlight the question of what level of geographic resolution is most appropriate for estimating the production of animal goods made possible by local plant resources. Here, in effect, SAR were considered as closed systems, with no circulation of crop or livestock resources (with the exception of live cattle). Finding a way to account for these movements, in and out, would represent a methodological improvement in assessing the capacity of a territory to supply animal feed consumption requirements. Another approach would be to conduct an analysis at larger levels of spatial organisation than the SAR, based on the organisation of agricultural production areas and/or on pedoclimatic and phytoecological characteristics.

4. Ecosystem services provided by agricultural ecosystems to society

Agricultural ecosystems provide other types of ecosystem services, the direct beneficiary of which is society as a whole (including farmers as members of society). These ES are primarily of two types: 1) regulating ES that help moderate phenomena detrimental to human wellbeing, such as climate change or the spread of contaminants into different environmental compartments; and 2) so-called "cultural" ES, which provide recreational, aesthetic, or spiritual benefits to society (Table 4-1).

This Chapter first describes ES from agricultural ecosystems that contribute to water quality regulation and to regulation of the global climate. Historically, agronomic research has focused more on assessing the negative environmental impacts of agricultural activities than on identifying and quantifying ES (see Chapter 1).

Then are presented the so-called "recreational" services provided by agricultural landscapes. These "services" are unique in that their CICES classification corresponds more to a typology of landscape uses and/or values than to ES in the sense adopted by EFESE-AE. Following a general discussion of the identification of "cultural" services, the goal of this Chapter is to develop a definition of recreational services that is compatible with EFESE-AE's overall analytical framework, identifying the processes or structural elements of agricultural ecosystems that help create suitable settings for the pursuit of recreational activities.

Table 4-1. Ecosystem services provided by agricultural ecosystems to society examined in EFESE-AE

Ecosystem service	Direct benefit(s) provided to society	Biophysical analysis
Soil stabilisation and erosion control	Fewer mudslide events	Quantified (see Chapter 2)
Storage and return of blue water	Blue water of sufficient quantity and quality for various uses (domestic, industrial, agricultural, recreational)	Quantified (see Chapter 2)
Natural attenuation of pesticides by soils		Methodological avenues
Regulation of water quality with respect to N, P and DOC		Partially quantified
Global climate regulation	Maintenance of current living conditions and human activities	Quantified
Recreational potential	Outdoor recreational activities with or without taking.	Partially quantified

 **Regulation of environmental conditions**

▌Agricultural ecosystem contributions to water quality regulation

Human use of water resources (for agricultural, industrial, domestic, recreational or cultural purposes) requires access to water in good ecological condition. Through its capacities for water retention and filtration, soil alters the physicochemical composition of rainwater infiltrating into and moving through it on its way to streams, lakes, aquifers and reservoirs (blue water).

The quality of water moving within and beyond the ecosystem is evaluated here in terms of the chemical substances and biological agents it may contain, increases of which can impact its suitability for various uses. Potential contaminants are diverse in nature, and may be biological (pathogens, viruses), organic (trace organic compounds or TOC, e.g., pharmaceutical substances) or mineral (trace mineral elements or TME) in nature. These compounds can enter the agricultural ecosystem through a variety of pathways, both natural (rain, atmospheric deposition) and anthropic (crop protection treatments, fertiliser applications, spreading of animal manures, sewage sludge, etc.). Nutrients essential to plant development (N, P, etc.) can also – in some chemical forms and at some concentrations – act as contaminants of aquatic environments.

Two major approaches are used in the scientific literature to evaluate regulating ES linked to water quality. The more common of these two approaches examines the mechanisms allowing for contaminant retention by the soil, also known as the ecosystem's capacity to "purify" (e.g., break down) or "filter" water (e.g., retain contaminants). Other research on these ES focuses on identifying and quantifying the contaminants present in water. The latter approach thus fits within traditional methods used by agronomists to evaluate the environmental impacts of agricultural practices. Contaminants exiting the agricultural ecosystem through water (runoff, lateral flow, percolation), through solid materials (erosion, exports of crop biomass), or in gaseous form (volatilisation) are considered as impacts of the agroecosystem on different environmental compartments (air and water). From the perspective of ES assessment, this second approach is less useful since it does not allow for an assessment of the effects of agricultural ecosystem characteristics and functioning on the regulation of material flows (Figure 4-1).

A given level of contaminants detected in drainage water can be the result of two different situations: (i) contaminant flow into the ecosystem is high, but is counterbalanced by a high capacity of the soil-plant system to prevent the movement of contaminants into drainage water (left); (ii) contaminant flow into the ecosystem is low, but the soil-plant system's capacity to filter or break down contaminants is also low.

Accordingly, regulating ES relating to water quality have been examined in EFESE-AE in terms of the capacity of the soil-plant system to limit contaminant movement into bodies of water. Based on the CICES classification system and the scientific literature, a distinction was made between ES regulating the quality of drained (or percolated) water with respect to N, P, and DOC, and ES for the natural attenuation of pesticides. Together, these two ES are responsible for mitigating two major sources of agricultural pollution: 1) nitrogenous and phosphorous fertilisation, and 2) crop protection products.

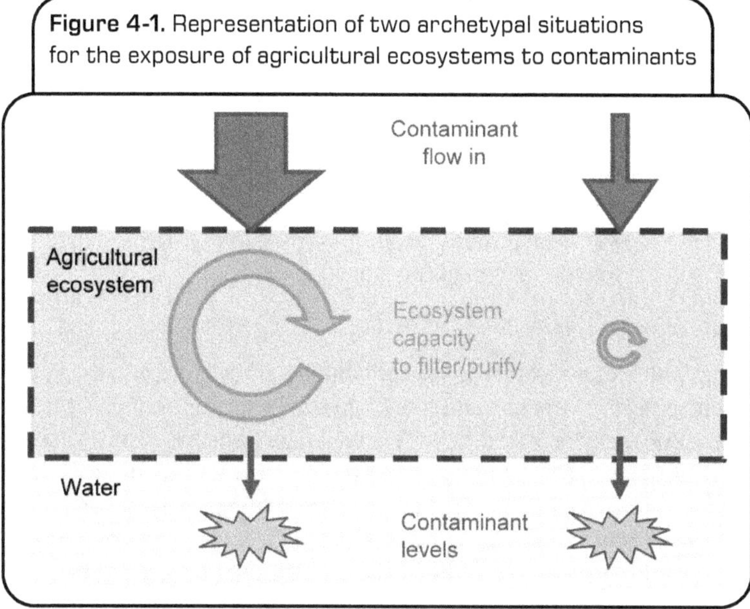

Figure 4-1. Representation of two archetypal situations for the exposure of agricultural ecosystems to contaminants

Figure 4-1. Representation of two archetypal situations for the exposure of agricultural ecosystems to contaminants

Biophysical determinants and external factors

Biophysical determinants

Once contaminants have entered the agricultural ecosystem, a number of different mechanisms are involved in limiting their movement into water bodies: absorption by plants followed by potential exportation from the system; retention in the soil through adsorption onto abiotic constituent materials (clays, organic matter); transformation through abiotic and/or biotic processes (pesticide breakdown, mineralisation/ organisation of N and P); and emission in gaseous form.

Regulation of water quality with respect to N, P, and DOC

Movement of P from agricultural ecosystems into waterways takes place primarily through the movement of soil particles by erosion (with the exception of a few very specific situations, such as very sandy soils). This is because orthophosphate ions bond strongly to the soil's solid phase. Since nitrate ions and DOC are soluble, these are primarily carried by water percolating through the rhizosphere through drainage. The movement of N and DOC beyond the root zone is thus strongly determined by the magnitude of vertical water movement. It should be noted that not all compounds carried beyond the root zone will necessarily find their way into water resources to become a contaminant.

The biophysical determinants of water quality regulation with respect to N, P, and DOC are the same as those for the ES "nutrient supply to crop plants", "storage and return of water to crop plants", and "soil stabilisation and erosion control" (see Chapter 2). Abiotic characteristics (nutrient levels in the soil, soil moisture, soil organic matter levels, etc.) and soil functioning play a key role, as do the nature and spatiotemporal distribution of vegetative covers during drainage periods or potential runoff events. In the case of pastured systems, animal dejections will impact the quantity of nutrients entering the ecosystem. The quantities of N distributed in a non-homogenous fashion over the soil surface by cows fed on pasture vary from 150 to 500 kg of N/ha, depending on animal stocking rates and the length of the grazing period.

Natural attenuation of pesticides by soils

Pesticides interact with biotic and abiotic soil components. They move into the liquid, solid, and gaseous phases of the soil and into biomass. Pesticides remaining in the liquid phase of the soil are considered to be "bioavailable," which is to say they can contribute to the pesticide exposure of soil-dwelling organisms. Local dissipation of pesticides, or the apparent diminution of pesticide concentrations in the soil, depends on three major types of processes associated with the biotic and abiotic components of the soil-plant system: i) retention in the soil, ii) transformation/breakdown, and iii) dispersion into other environmental compartments (plants, animals, water, the atmosphere).

The physicochemical properties of the soil (particularly soil texture, soil organic matter level, soil temperature) and of the specific pesticides involved determine the degree to which pesticides adsorb onto soil constituents. Adsorption can lead to the formation of bonded residues, the precise nature and fate of which are unknown. Abiotic characteristics also determine whether pesticides can be broken down by photolysis or hydrolysis, volatilised, or transferred into water bodies.

Microbial biodegradation is the most important process involved in pesticide breakdown in soils, and will depend on the abundance and functional diversity of soil microorganisms. Pesticide breakdown can be partial (co-metabolism), leading to an accumulation of pesticide metabolites that can sometimes be more persistent and/or more toxic than the parent molecule; or total (metabolism) leading to a mineralisation of the pesticide into, for example, CO_2 and NH_4. Soil pH is a major edaphic parameter determining pesticide biodegradation. Biotic interactions in soils are also known to influence the microbial biodegradation of pesticides. For example, the intensity of biodegradation of several herbicides can vary depending on the type of planted vegetative cover, or again as a function of micro-environments created by soil fauna (e.g., worm galleries). Finally, the nature and spatiotemporal distribution of plant cover will determine the absorption of some amount of pesticides into plant roots.

The natural attenuation of pesticides by soils is not infallible. Adsorption is a reversible process, and thus can contribute to delayed pollution phenomena, the dynamics of which are poorly understood. Pesticide breakdown can also lead to the accumulation of metabolites which themselves become contaminants. As a result, the soil can be a source of secondary contamination phenomena, the behavior of which is difficult to predict.

External factors

Regulation of water quality with respect to N, P, and DOC

Fertilisation (mineral or organic) affects the quantity and chemical form of nutrients entering the soil-plant system. The dynamics and intensity of N mineralisation are likewise influenced by the management of crop residues (exportation of residues at physiological maturity, mulching at the soil surface, incorporation) and by climate (influence of temperature and solar radiation on plant development).

Water movement is also influenced by soil tillage practices (impact on available water reserve), irrigation, and climate (precipitation, evapotranspiration).

Natural attenuation of pesticides by soils

Contamination of agricultural soils by means of pesticide applications to crops will depend on the type of product used, on the application method, and on the timing of pesticide applications relative to the crop growth stage. All of these factors will influence the extent to which pesticides are intercepted by crop foliage vs. falling directly on the soil.

Numerous climatic parameters, including soil temperature and humidity as well as cultural practices, modulate the biodegradation of pesticides. The repeated application of crop protection treatments exerts selection pressure on soil microbial populations, making total metabolism more effective through processes of adaptation. Soil tillage also causes profound changes in soil structure and soil porosity, which can affect hydraulic connectivity and, as a result, alter the distribution and presence of soil microorganisms. Organic matter inputs, whether of plant or animal origin (straw, manures), foster the growth and activity of microbial biomass developing in the detritus layer. Finally, liming can influence pesticide breakdown through its effects on soil pH.

Assessment methods for the regulation of water quality

No established methodology exists for assessing the capacity of soils to regulate pesticide movement into drainage water. This is most likely due to the complexity of the interacting abiotic and biotic processes involved. Currently available data and assessment methods only offer a partial understanding of this ES. In general, these methods either allow for the quantification of specific pesticide residues in soils and/or water (a measure of the environmental impact of crop protection practices), or offer an assessment of some of the processes involved in pesticide breakdown.

Within the scientific literature on ES for water quality regulation, the assessment of ES relating to the soil's capacity to inhibit the loss of nutrients into water resources seeks to quantify the "ecological work" performed by the ecosystem. The MAES program suggests estimating the quantity of nutrients "retained" by the system over a given period of time, measured through nutrient levels in the soil. Other researchers have calculated the difference between the level of ecological pressure (e.g., the volume of N inputs) and environmental quality (e.g., the amount of leached N). Still other authors have sought to create an index of soil vulnerability to leaching, although this approach does not account for the agricultural ecosystem's capacity to retain nutrients through uptake by vegetative covers before and/or during periods of drainage and runoff. Some tools employed by agronomists do allow for consideration of this "soil-plant" dimension within ES assessment. In addition to field measurements – which are expensive to undertake on a large scale – models such as STICS and PaSim, which can dynamically simulate water and N balances according to different management practices and pedoclimatic conditions (see Box 2-1), are used for this purpose.

Level of ES provision

Given the current availability of data and assessment tools (particularly modeling tools), water quality regulation with respect to N was the only aspect of the ES for water quality regulation this study sought to quantify. In addition, the study proposed some methodological avenues for assessing i) the natural attenuation of pesticides by soils and ii) the regulation of drainage water quality with respect to P and DOC (see below).

Assessment method

The simulation tool developed specifically for EFESE-AE was used to estimate the effective level of regulation provided by agricultural ecosystems for drained water quality with respect to N. The indicator that was calculated corresponds to the annual average quantity of non-leached N, that is, the amount of N "retained" by the soil-plant system or lost in the form of gaseous emissions. Since the assessment was focused on the regulation of N movement into water, gaseous emissions of N in the form of N_2, N_2O, and ammonia (NH_3) were accounted for but are not specifically examined here. These gaseous emissions correspond to negative impacts of agroecosystems (N_2O and NH_3 emissions) and are linked to the level of the ES for global climate regulation (reduction of N_2O into N_2), but from the point of view of water quality regulation, they amount to N that is not leached. The analysis of an ES for air quality regulation – not attempted here – would allow for the assessment of ecosystem effects on the fluxes of NH_3. Viewed more broadly, this situation highlights the fact that the indicator chosen for each ES will relate directly to the flows of matter or energy of interest to that particular assessment.

The indicator for the ES for water quality regulation with respect to N was calculated as follows:

non-leached N = inputs of mineral N (fertilisation, net mineralisation) – leached N

This quantity can also be expressed as a percentage of N inputs (quantity of non-leached N divided by mineral N inputs). Both indicators are evaluated while accounting for the supply of inputs (according to "current" agricultural practices).

"Current systems" simulations, in which management systems are assumed to correspond to current, prevailing management methods for agroecosystems (in terms of mineral fertilisation, crop residue management, and irrigation for maize), made it possible to estimate the level (in absolute or relative terms) of the ES effectively supplied by agricultural ecosystems.

In addition, the two indicators were calculated using two alternative simulation scenarios:
  • the effect of cover crops was tested by using an alternative "no cover crop" simulation scenario for PCU located in Nitrate Vulnerable Zones (all other practices remaining the same);
  • the effect of irrigation was tested by using an alternative "no irrigation" simulation scenario for PCU including maize crops ordinarily irrigated (all other practices remaining the same as in the reference simulations).

Results

Only those results obtained for major field crops thanks to the STICS model are presented here. The two indicators were interpreted with regard to various characteristics of the simulated agricultural ecosystems: soil type, available water reserve, climate type, rotation type,[34] etc.

The annual average quantity of non-leached mineral N varies from 122 kg N/ha/yr to 363 (average of 242 kg N/ha/yr), or between 55% and 100% (average of 86%) of the N mineralised by the ecosystem and/or imported through nitrogenous fertilisation (Figure 4-2).

The highest absolute levels of the ES (>264 kg non-leached N/ha/yr) are primarily found in southwestern France (Landes, the foothills of the Atlantic Pyrenees, along the Garonne); in the heart of the Parisian basin; in Flanders; in parts of Alsace; and in parts of the Rhône valley. It is possible that these regions are characterised by higher levels of N uptake by standing crops (for example, irrigated maize; see below).

---

34. Not all combinations of [indicator X variable] could be examined in EFESE-AE.

Some regions with a lower absolute level of the ES are nevertheless associated with a high relative level of the ES. This is the case, for example, with the middle Garonne basin, western France and the Lorraine plateau, where approximately 90-100% of N is not leached. The divergence of results between the two indicators may be explained by the fact that a low level of non-leached N can result from a low quantity of mineral N present in the ecosystem – for example, as a result of low levels of external inputs (on the order of 50 kg N/ha for sunflower).

Figure 4-2. Estimated annual average quantity (A) and percentage (B) of non-leached N in cropping systems managed with observed agricultural practices

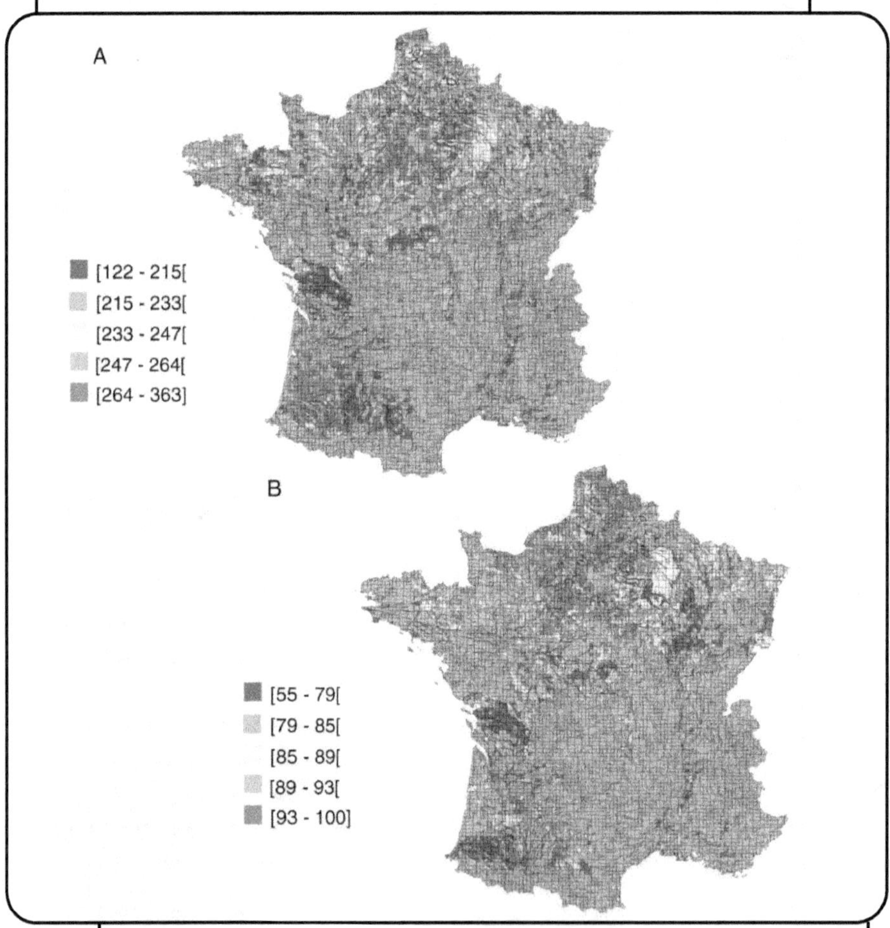

A: Quantity of non-leached N (in kg N/ha/yr); B: Proportion of non-leached N (in %). Spatial resolution: PCU. PCU in gray (including Corsica): no "major field crop" simulations. PCU in white: excluded from the analysis.

Finally, we find lower levels, both in quantity and as a percentage, in the southern Rhône Valley, in Berry, and in Poitou. The isolated red points on the map could represent specific situations characterised by low yield potentials and excessive (simulated) inputs of mineral N. These results should be examined in more detail in order to understand the factors at work and to confirm or reject this hypothesis.

The quantity of non-leached N is strongly positively correlated with the quantity of N taken up by the plant cover. This positive relationship also exists – although it is less strong – for the indicator of the relative level of the ES.

The indicator of the absolute level of the ES is also positively correlated with the level of mineral N fertilisation and the quantity of mineralised N, and seems to be higher where the UR is higher. These correlations may be explained by the positive relationship between N uptake by the plant cover and crop growth, the latter of which is favored by fertiliser inputs, high rates of mineralisation, and soils with a high UR. In addition, a large majority of mineral N is "retained" by the soil and the plant cover in current major field crop systems, with only a small part lost in the form of gaseous emissions.

Finally, the indicators are not correlated or are only weakly correlated with drainage intensity: only a weak negative correlation is shown between the percentage of N not leached and the quantity of drainage water. This is a surprising result that should be examined further, since it has been demonstrated in the scientific literature that drainage strongly affects levels of leached N.

Effect of crop rotation type on the level of ES supply

At the national level (France), no correlation was found between the indicators of the level of ES and various rotation characteristics: rotation length, percentage of cereals in the rotation, percentage of maize, percentage of winter crops, percentage of legumes. On the other hand, comparison of the "with" vs. "without" cover crops simulations reveals an overall positive effect of maintaining soil cover between crops on the absolute and relative levels of the ES. This is true regardless of climate type, soil texture, or nature of the crop rotation. This result is in keeping with the fact that the level of ES supply is strongly linked to N uptake by the plant cover, with the cover crops serving to use N in between growing periods for the commercial crops. This effect is limited, however, by the low frequency of cover crops use within the simulated crop rotations: on average, cover crops absorb 13 kg N/ha/yr.

Effect of irrigation on the level of ES supply

All other things being equal, comparison of the "with" vs. "without" irrigation simulations for PCU with maize crops that are ordinarily irrigated shows that irrigation improves ES levels in both absolute and relative terms. This positive effect is largely due to an increase in the quantity of N absorbed by the commercial crop when it is irrigated. A determining factor in this result is probably the failure to adjust nitrogenous fertilisation rates to match the yield potentials achievable without irrigation. Simulated fertilisation rates were the same for the two situations, whereas the potential for N uptake is very different when water supplies are limited. It is likely that the irrigation effect would be lower (although still positive) if nitrogenous fertilisation rates were reduced.

Perspectives for improvement

It would be interesting, first, to extend the analysis of the effect of agricultural practices on ES levels by refining the simulations. As with the assessment of the supply of mineral N to crop plants (Chapter 2), fertilisation practices could only be represented here in very general terms (at the regional level) due to a lack of more detailed data on specific agricultural practices. In addition to adjusting simulated fertilisation rates to match yield potentials for each pedoclimatic context, closer examination of the effect of the type of nitrogenous fertilisation (mineral vs. organic) on ES levels is needed. Again, as noted above, the limited observed effect of cover crops on the quantity and percentage of non-leached N may be the result of low levels of cover crop use in the simulations. The effects of a more systematic use of cover crops, with longer coverage periods, should be examined. Finally, the analysis only looked in a limited way at the impacts of crop residue management and soil tillage strategies, factors that merit special attention in future work.

As a more long-term prospect, it would be useful to describe the temporal dynamics of ES levels as a function of changes in soil organic matter rates as influenced by simulated cropping systems. It should be possible to observe trends toward an increase or a reduction in ES levels in association with this dynamic. More broadly speaking, the effect of so-called "alternative" cropping systems, such as those based on a combination of no-till, permanent soil cover and diversified rotations, deserves to be explored. To do so, it will be necessary to develop a better understanding of the processes involved, assess the capacity of existing simulation models to represent these cropping systems, and, if necessary, adapt the models in order to evaluate the performance of these systems across a wide range of pedoclimatic situations.

The assessment of this ES should be completed by integrating the contaminants ignored here, particularly pesticides, P and DOC.

Additional research is needed to gather data on pesticide biodegradation in different pedoclimates and to develop predictive models of the attenuation potential of different agricultural soils. Recent work in microbial ecology use molecular techniques based on the direct extraction of nucleic acids from soil to quantify the abundance and activity of pesticide-degrading microbial populations. They suggest that these nucleic acids could constitute bio-indicators accounting for exposure to pesticides and the potential for chemical mitigation. Although such methods have been standardised with the International Organisation for Standardisation, they have yet to be implemented on a large scale. A forthcoming European Directive on soil protection may require *a posteriori* impact assessments of pesticides on soil microorganisms, which could lead to the systematic use of such analyses. The European Food Safety Authority (EFSA) has also published a number of scientific opinions on the *a priori* assessment of the environmental risks posed by pesticides, which has encouraged scientists to develop microbial bio-indicators of ES provided by soils, including pesticide attenuation.

The quantity of P not lost by erosion could potentially be used as an indicator of water quality regulation with respect to P. This indicator could be calculated using a methodology that would bring together the spatial distribution of P stocks in the arable layer of agricultural soils in France (existing cartography carried out by Delmas *et al.*, 2015[35]) with a mapping of soil erosion risks (that produced for the assessment of the ES for soil stabilisation and erosion control – see Chapter 2). The indicator could be further refined to account for inputs of phosphorous fertilisers, resulting in an assessment of the effects of P fertilisation on ES supply.

A crop model such as EPIC[36], which simulates DOC fluxes in the soil and beyond the root zone, could be used to provide an initial assessment of the capacity of agricultural ecosystems to limit losses of DOC into water percolating below the rhizosphere. This would require a significant effort of model calibration for agropedoclimatic conditions in France.

❚ Contribution of agricultural ecosystems to climate regulation

The increase in temperatures at the Earth's surface since the beginning of the industrial era, a manifestation of climate change, depends on the concentration of GHG in the atmosphere. Climate regulation corresponds to the mechanisms helping to limit the increase in average global temperatures, and thus implies a reduction in global GHG emissions. In France, the agricultural sector accounts for 19% of GHG emissions.[37] The attenuation of GHG emissions linked to agriculture accordingly represents a major objective within the climate change policy framework adopted by the French government (with a goal of reducing GHG emissions by 75% relative to 1990 levels by the year 2050). In the agricultural sector, reductions could be achieved through changes in the overall agricultural system, changes in agricultural practices, or again by favoring certain ecosystem processes, including C storage in soils and in woody biomass.

Most research on GHG emissions from agricultural ecosystems does not make use of the ES concept. A central methodological challenge in examining the interrelationships between agroecosystem functioning, GHG emissions, and climate regulation is to distinguish between assessments of the impact of agricultural practices on GHG emissions (notably the use of nitrogenous fertilisers) and assessments of the ES provided by agricultural ecosystems that can help regulate GHG emissions.

---

35. Delmas, M., Saby, N., Arrouays, D., Dupas, R., Lemercier, B., Pellerin, S., Gascuel-Odoux, C., 2015. Explaining and mapping total P content in French topsoils. Soil Use Manag. 31, 259–269. doi:10.1111/sum.12192
36. http://epicapex.tamu.edu/epic/
37. Including CO_2 emissions related to fossil fuel use by the agricultural sector.

In EFESE-AE, the ES of global climate regulation is defined as all processes relating to i) C storage in the soil and in woody biomass directly associated with agricultural ecosystems (hedges around field perimeters, trees within or at the edges of fields), and ii) the attenuation of N_2O and CH_4 emissions, the two principal GHG associated with agriculture. Note that flows of GHG linked to energy consumption and to changes in land use are beyond the scope of the assessment.

Biophysical determinants and external factors

Biophysical determinants

C storage

While a significant fraction of the organic C introduced into the soil is rapidly mineralised by microbial biomass, some is incorporated into the soil in a durable manner, with residence times that can extend from several years to several centuries. Planned and associated biodiversity plays a key role in C storage in the soil. Thus, the nature and the spatiotemporal distribution of plant covers (grasslands, crops) as well as the harvesting of biomass, the management of crop residues, and the return of animal manures to the soil by grazing animals affect the quantity of C introduced into the soil, its biochemical nature, and its degree of biodegradability.

Plant cover characteristics also determine the intensity of heterotrophic respiration from microbial activity, which constitutes another path for the movement of C into the atmosphere in the form of CO_2. Soil fauna and microbial biomass are involved in transformations of the C incorporated into the soil (consumption, redistribution, mineralisation, etc.).

With respect to abiotic determinants, soil structure and soil texture determine soil temperature and soil oxygenation and moisture levels, which, combined with soil composition characteristics (e.g., allophanes, carbonates, nutrients), determine the activity of soil organisms and thus the rate of C mineralisation and processes of organic C stabilisation. Finally, ES for soil stabilisation and erosion control contribute to the conservation of stored C by protecting the soil surface layers, which usually have the highest levels of C (see Chapter 2).

C storage in woody biomass associated with agricultural ecosystems comes from CO_2 removed from the atmosphere by photosynthesis. Its rates will be determined by the nature and spatiotemporal distribution of woody vegetation (isolated trees, thickets, different types of hedges, etc.). Note that nearly 80% of field parcels in the French LPIS (2012) include woody plants within their defined spatial boundaries (not including vineyards and orchards).

N_2O and CH_4 fluxes

CO_2 emissions account for 10% of GHG emissions from the agricultural sector, while N_2O and CH_4 account for 50% and 40% of CO_2 equivalent (CO_2e), respectively.

Based on current research, it is difficult to distinguish between biophysical factors relating to N_2O emissions and biophysical factors relating to the soil's capacity to regulate N_2O emissions. N_2O is produced by microbial transformations of N in the soil and in animal effluents (nitrification[38] and denitrification[39]). Emissions are primarily linked to the use of nitrogenous fertilisers, both mineral and organic. In addition to the bacterial communities present, which contribute to the denitrification potential, the principal physicochemical factors that have been identified as determining emissions levels are nitrate and ammonium concentrations, the availability of organic C, and the degree of soil water saturation (which determines the relative importance of nitrification vs. denitrification), soil temperature (which influences the activity of microorganisms), and pH (which particularly affects the soil's capacity to reduce N_2O into N_2).

CH_4 production results from the activity of a methanogenic microflora active during the fermentation of organic matter under anaerobic conditions. In France, nearly all CH_4 emissions from agricultural ecosystems are enteric emissions from ruminants, the result of microorganism activity in their digestive systems. Apart from the nature and activity level of the bacterial microflora in the rumen as determined by animal diet, the primary biophysical factors affecting CH_4 emissions are animal species, genotype, and age. CH_4 emissions from livestock effluents are for the most part produced outside the agricultural ecosystem proper (i.e., in livestock buildings or storage facilities) and so were not considered here.

External factors

Independently of climate, which affects the intensity of the processes at work in the N and C cycles, certain agricultural practices have an effect on C storage in soils and woody biomass and on the fluxes of N_2O and CH_4.

C storage

Fertilisation, irrigation, residue management at harvest (export, mulch, incorporation), grassland management practices (haying, mowing), and organic amendment inputs (type, quantity, application method) are all involved in determining the dynamics and rate of return of C to the soil. Soil tillage can influence C storage through its impact on the incorporation of organic materials and the rate of C mineralisation in the soil. Methods for the maintenance or exploitation of woody growth associated with agricultural ecosystems (hedges, copses, etc.) determine their growth and thus their associated C storage dynamics.

N_2O and CH_4 fluxes

Use of mineral and organic nitrogenous fertilisers affect soil nitrate and ammonium concentrations. Management of soil pH by liming or the use of other amendments will

---

38. Biological oxidation of ammonium into nitrite and then into nitrate: $NH_3 \rightarrow NO_2^- \rightarrow NO_3^-$.
39. Successive reduction of oxidised soluble forms of N into gaseous components: $NO_3^- \rightarrow NO_2^- \rightarrow NO \rightarrow N_2O \rightarrow N_2$.

also influence N_2O emissions, since these can be reduced by an increase in soil pH. Soil tillage and the management of soil moisture levels (drainage, irrigation) influence the physicochemical conditions that affect N transformation in the soil. Finally, the feeding strategies for ruminant livestock and methods for spreading livestock effluents strongly influence CH_4 emissions.

Level of ES provision

Most research on ES for global climate regulation focus on ecosystems that are relatively untouched by human activity, and use C storage in the soil or in woody vegetation as indicators of the ES level. Studies interested in the agricultural sector tend to focus more on the negative climate impacts of agroecosystems than on agricultural ecosystems' capacity to help regulate climate. Whereas early research would often look at just one of the three GHG (CO_2, N_2O and CH_4), a growing number of references now address the overall GHG balance of agroecosystems; that is, annual net emissions of the three GHG weighted according to their respective global warming potential (GWP). Approaches vary in terms of how the boundaries of the system to be examined are defined and thus the relevant spatio-temporal scale. Two broad types of approach may be identified: i) "source-sink" approaches, which quantify flows of GHG between the atmosphere and the ecosystem under consideration at the level of the field unit, group of fields, geographic area, or country; ii) "life-cycle analysis" (LCA) approaches, which assess the environmental impacts (particularly the "carbon footprint") of a system supplying a specific product or service, from extraction of the primary materials, through fabrication and delivery, through disposal at the end of its useful life.

The "source-sink" approach is more suitable for ES analysis since it focuses on ecosystem functioning. This is the approach adopted by the MAES program, in which research is based on indicators of the stock or flow of C reported per unit of ecosystem land area. LCA approaches, moreover, consider the totality of the socio-ecological system, of which the agroecosystem is only a part. LCA thus cannot be applied to agricultural ecosystems alone.

Assessment method

The ES has two components: i) C storage in the soil, and ii) the attenuation of N_2O and CH_4 emissions. Soils' capacity to convert N_2O into N_2 varies widely depending on soil type, the soil microbial communities present, and the physicochemical conditions that influence soil microbial activity. Identification of the biological and physicochemical factors underlying this variability at different scales is one focus of current research. Available data do not allow for a mapping of this component of the ES for climate regulation at the level of mainland France, however. Assessment methods for an ES for the regulation of enteric CH_4 emissions also have yet to be developed. Accordingly, only the "C storage" component of the ES was quantified in EFESE-AE.

In keeping with existing work, two indicators were calculated: i) the quantity of C stored in soil organic matter and woody biomass, which measures a quantity of C durably removed from the atmosphere due to the existence of the ecosystem; ii) the annual variation in C stocks, which measures the current ecosystem contribution to the reduction of CO_2 concentrations in the atmosphere. These two indicators correspond to two elements contributing to the "C storage" component of the ES for global climate regulation: i) maintenance of the current stock of C already present in the ecosystem, and ii) additional storage of atmospheric C over time.

The stock of C in soils was calculated at a detailed level of resolution using the results from previous work on soil C levels and associated total C stocks. C stored in woody biomass was estimated by multiplying the quantity of C stored per unit of land area of woody growth[40] by the land area occupied by woody growth in field parcels inventoried in the French LPIS (vegetation layer of the IGN (National Institute of Geographic and Forest Information) TOPO® database). With the land-use information contained in the LPIS, these two indicators could be calculated at the level of the cropping unit and then aggregated at the level of the PCU.

The change in the annual average soil organic C levels was calculated using the STICS and PaSim simulation tools. This corresponds to the annual average variation in levels of C stored in the soil between the first day and the last day of the simulation (simulated over 30 years – cf. Box 2-1). As for the assessment of the regulation of drained water quality with respect to N, "current systems" simulations were used to estimate the level of ES effectively supplied by the agricultural ecosystems given the addition of inputs made over the reference period. The effect of cover crops on the level of ES supply was tested using an alternative simulation scenario "without cover crops" for PCU located in Nitrate Vulnerable Zones (other practices remaining unchanged).

Current C stocks in agricultural ecosystems

Total C stocks in agricultural ecosystems, including both C stored in the soil and C stored in woody growth within field boundaries, average 71 t C/ha (Figure 4-4). Figure 4-5 shows the geographic distribution of C stored in woody material as a percentage of total stored C.

On average, the stock of C in the 0-30 cm soil horizon is 59 t C/ha in major field crops and 76 t C/ha in grasslands. These values are similar to those found in the scientific literature. C stocks in the 0-1 m soil horizon (not shown here) are a little less than twice those observed for the 0-30 cm horizon. These results show the combined effect of pedoclimate and land use on C stocks. In keeping with other findings reported in the literature, the highest stocks were observed in high-altitude areas (Alps, Pyrenees, Massif Central, Jura, Vosges) and/or in grassland areas (Brittany, Lower Normandy). Higher levels in mountainous areas result from the combined effect of high-altitude climate regimes (low temperatures and higher precipitation during anoxic periods less favorable to C mineralisation in the soil) and land use (dominance of permanent grassland). Conversely, the lowest levels were observed in lowland areas and major

---

40. Data from the Climagri® tool developed by the Agency for the Environment and Energy Management http://www.ademe.fr/expertises/produire-autrement/production-agricole/passer-a-laction/dossier/levaluation-environnementale-agriculture/loutil-climagri

field crop zones (the Parisian basin, the Aquitaine basin, the Saône Valley, the Rhône Valley, Alsace, the Limagne plain). Note that the higher C stocks in major field crop areas in Brittany, Charente-Maritime, Lorraine, and the Barrois plateau can be explained by historical land use (soils previously in grassland in Brittany and Charentes) and/ or by soil type and climate (clayey soils in the Marais Poitevin, clayey soils and cold climate on the eastern edge of the Parisian basin).

By extrapolating these results to all land areas in major field crops (15.8 Mha) and in grasslands (9.7 Mha) – as estimated based on the French LPIS – the 0-30 cm soil horizon in these two land-use categories represents a total C stock on the order of 1.75 billion metric tons, or 47% of total stored C in French soils.[41] This is the equivalent of 16 years of GHG emissions in France for all sectors combined, or twice that if one considers the soil horizon from 0-1 m. Despite having a lower level of stored C per unit of land area than grasslands, major field crop agricultural ecosystems represent a higher total level of stored C due to their greater land area.

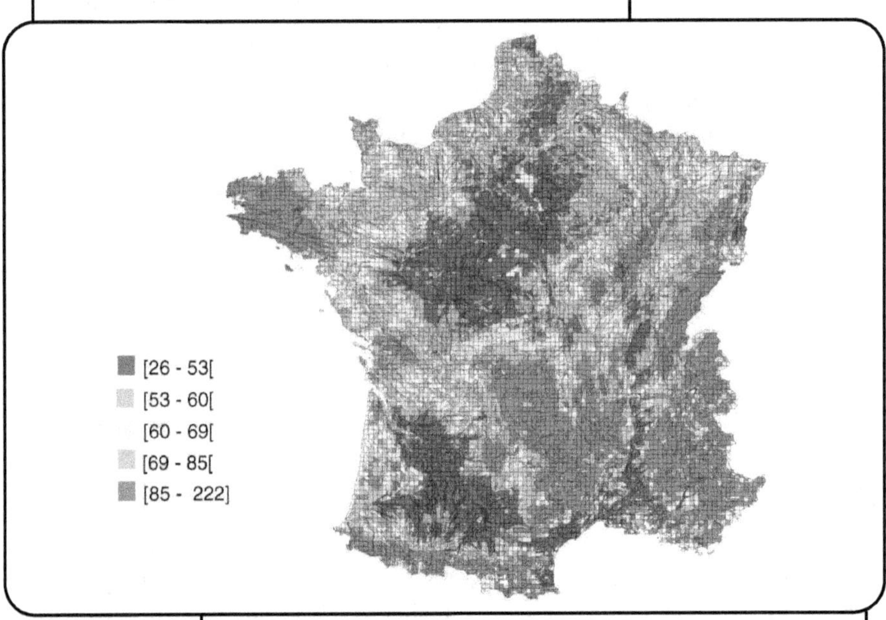

Figure 4-3. Total C stored in agricultural ecosystems (soil to a depth of 30 cm + woody material) in t C /ha

Spatial resolution: PCU. PCU in white (including Corsica): not estimated. Value classes correspond to quintiles.

---

41. Estimated at 3.725 Pg according to Meersmans, J. *et al.* (2012). A high-resolution map of French soil organic carbon. Agronomy for Sustainable Development, 32(4), pp. 841–851. doi: 10.1007/s13593-012-0086-9

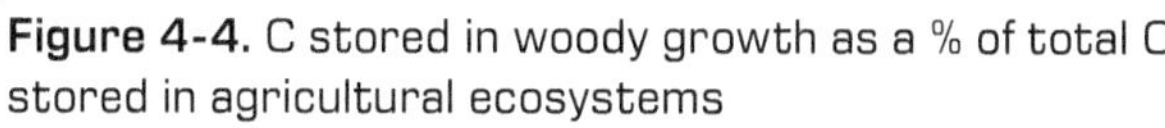

Figure 4-4. C stored in woody growth as a % of total C stored in agricultural ecosystems

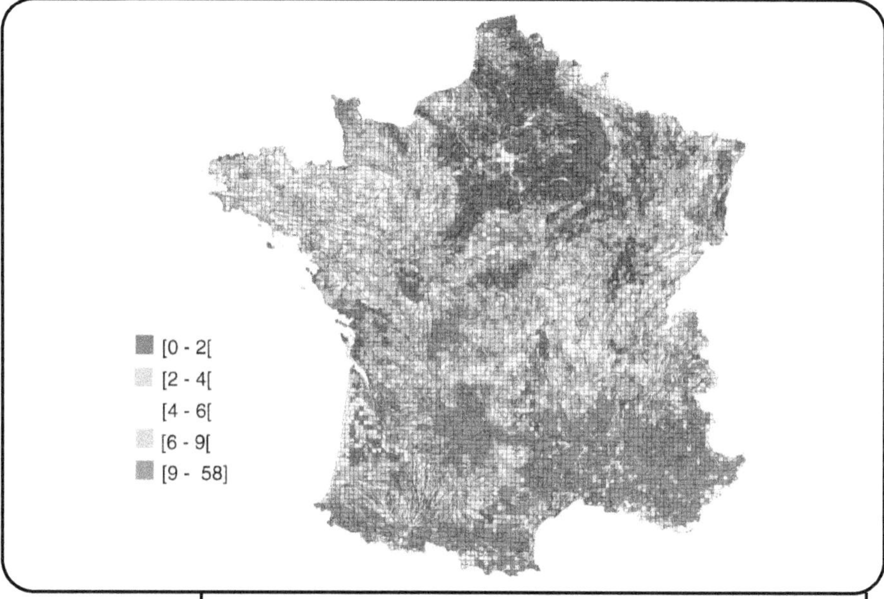

Spatial resolution: PCU. PCU in white (including Corsica): not estimated.
Value classes correspond to quintiles.

Figure 4-4 shows the relative importance of C stored in the soil vs. total C stored in agricultural ecosystems. Note that the C stock associated with woody growth represents on average just 7% of total C stored in the ecosystem (less than 5% if one considers the 0-1 m soil horizon), and is very rarely more than 20%. This percentage is particularly low in the Parisian basin (with the exception of the western edge), despite lower soil C levels, given the small amount of land area allocated to woody growth. It is higher in the Mediterranean area, in mountainous areas, and in the Sologne. In absolute terms, C stocks associated with woody growth vary from 0 to 8 t C/ha, with an average of 5 t C/ha. By design, the geographic distribution of indicator values is directly linked to the geographic distribution of woody materials: the lowest levels are found in major field crop areas, notably in the Parisian basin where woody formations are rare. Conversely, mountainous areas (and, to a lesser extent, the bocage of western France) show the highest levels of stored C. It should be noted, however, that although woody formations contribute little to the total amount of stored C in agricultural ecosystems, their conservation is nevertheless highly important given the importance of semi-natural areas in the supply of many other ES (regulation of pest species, pollination, etc.).

Annual change in organic C stored in soils in major field crop ecosystems

Figure 4-5 shows the annual average variation in C stored in soils in major field crop systems as simulated with STICS, using current agricultural practices. For the most part, simulated annual storage rates range from −0.5% to +0.4% (or from -5 to +4 per thousand), with an average loss of -0.03% per year (or -0.3 per thousand).

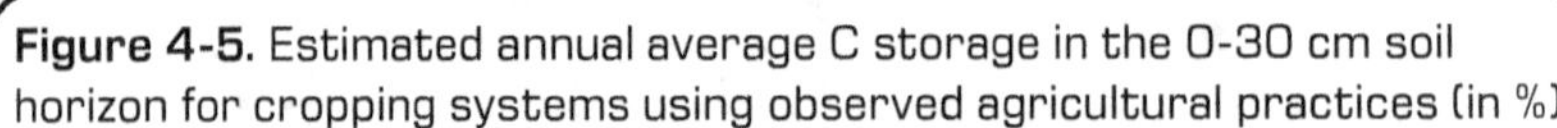

Figure 4-5. Estimated annual average C storage in the 0-30 cm soil horizon for cropping systems using observed agricultural practices (in %)

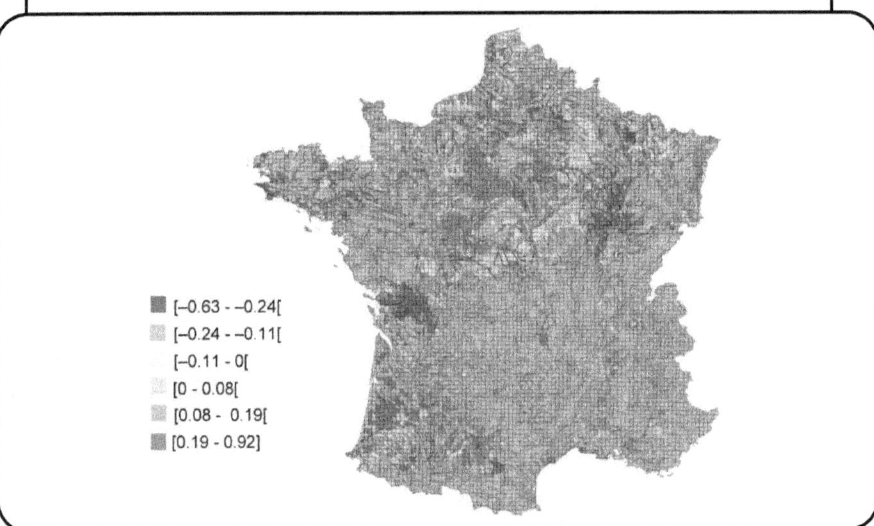

Average annual relative change in stored C. Spatial resolution: PCU. PCU in gray (including Corsica): no "major field crop" simulation. PCU in white: excluded from the analysis. The three negative classes and the three positive classes are of equal size.

Locations showing an annual loss of C (shown in red) correspond to areas where initial C levels were high (Brittany, Charente-Maritime, Lorraine, and the Barrois plateau) (see Figure 4-3). This result suggests that the cropping systems currently practiced in these areas will not maintain C stocks in the soil at their current level. In Brittany, where stored C in arable soils is high, the simulations show a trend towards loss despite regular organic fertiliser inputs associated with livestock production. Given the assumptions made in the simulation scenario, these inputs do not appear to fully compensate for the dynamics of C loss linked to low levels of incorporated crop residues (maize harvested as silage, straw exported to other areas) and climatic conditions that favor mineralisation. Conversely, in major field crop areas characterised by low initial levels of stored C, the simulations suggest that current cropping systems are capable of maintaining or even slightly increasing stored C (Parisian basin, mid-Garonne basin, Alsace). This result is in line with other research, and can be understood as resulting from simulated practices appropriate to areas located in Nitrate Vulnerable Zones and where livestock production is rare: incorporation of cereal straw/maize stover, use of cover crops.

Note that in major field crop regions, the increase in simulated C stocks is mostly below 2 per thousand (0.2%), and is very rarely higher than 3 per thousand (0.3%).

An examination of the correlation between the ES indicator level and the soil type shows that C storage is higher in clayey soils than in sandy soils, in keeping with what we know about the biophysical factors and external factors that control soil C dynamics. This "soil effect," combined with an oceanic climate favorable to mineralisation, could also explain the reduction in simulated stocks shown in the Landes. C storage tends to increase as organic amendment inputs rise, but the relationship is loose, probably due to the multiplicity of factors that influence the fate of these materials (e.g., current C levels, climate).

Comparing simulations with and without cover crops confirms the positive effect of the latter on C storage: on average across all simulations, the loss of C is reduced by 38 kg C/ha/yr in the presence of cover crops. As with the ES for water quality regulation with respect to N, however, this relatively limited effect of cover crops on C storage is probably attributable to the limited use of cover crops within the simulations and the relatively short periods of time fields were simulated as being "in cover crops."

Perspectives for improvement

This assessment shows that a spatialised quantification of the potential for additional C storage across pedoclimatic and agronomic contexts is possible. While the technical levers involved are relatively well understood, the strategies to be prioritised in different contexts and the possibilities of combining multiple levers require further exploration. A more detailed analysis of the simulation results could contribute to this end through an analysis of the effect of crop rotations, fertiliser type, cover crop strategies, management of woody material, etc. on the C dynamics of different cropping systems. In parallel, continued research should be pursued on the processes responsible for N_2O emissions; this assessment confirms the importance of N_2O emissions within the GHG budget of agricultural ecosystems. Note that a pedotransfer function[42] allowing for the mapping, at the national level, of soils' capacity to reduce N_2O into N_2 is currently being developed by Inra.

Systemic and large-scale approaches should complement those implemented at the local level. Other cropping system configurations should be simulated to test the effects on C storage at the territorial level (e.g., the re-association of crop and livestock systems or the increased recycling of organic/green wastes from urban areas). Given the central role of organic C in multiple soil properties and in other ES supplied by soils, the effects of increased soil C levels on these other ecosystem services should also be examined.

---

42. Tool based on statistical relationships allowing certain soil properties (difficult to measure directly) to be estimated from other well-known ones.

Contribution to the recreational potential of the landscape

The definition of ecosystem "cultural" services

CICES, in its definition of "cultural services," focuses on the immaterial benefits human beings receive from contact with flora, fauna, or ecosystems in general. The CICES classification differentiates between two broad sub-categories of cultural services according to the nature of the "cultural" interactions human beings maintain with the biophysical components of ecosystems and landscapes: i) "spiritual and symbolic" interactions, and ii) "physical and intellectual" interactions (Table 4-2).

Table 4-2. The nature of "cultural services" according to CICES

	CICES typology	Examples
"Spiritual and symbolic" interactions with the biophysical components of ecosystems and landscapes	Symbolic value or sacred value	Emblematic plants and animals, sites with spiritual associations, etc.
	Existence value or transmission value	Desire to preserve biodiversity and satisfaction in knowing that it exists
"Physical and intellectual" interactions with the biophysical components of ecosystems and landscapes	Physical interactions in situ	Recreational activities in situ: walking, hunting and fishing, nature observation, etc.
	Intellectual and ex situ interactions	The ecosystem as a subject for experimentation, artistic representation, etc.

This definition is out of sync with the position that distinguishes between ES proper and the benefits provided by ES (see Chapter 1). This ambiguity was repeatedly pointed out during the public consultations on this classification that were organised by the European Agency for the Environment. The "cultural services" category is the category of ES most frequently cited as problematic, suggesting the difficulty of defining and describing this type of ES. The comments received point to a need to review this terminology and set of definitions, to eliminate all terms that refer to a benefit rather than to a service (for example, "recreation" refers to how society uses the ecosystem), and to approach these services from the perspective of the biophysical elements that enable or make possible an improvement in human wellbeing. Note that the French EFESE program takes a first step toward the clarification of this category of services by choosing not to consider spiritual and symbolic interactions as ES. The biophysical components that support such interactions are recognised as an aspect of natural heritage.

In keeping with the recommendations made by CICES during these consultations, a distinction was made in EFESE-AE between biophysical components associated with intellectual or symbolic representations and biophysical components that serve as a support for physical interactions. Priority was given to the latter as being more readily identifiable given the current state of research and currently available data. The following sections thus propose a specification and some initial avenues for the assessment of ES linked to the development of in situ recreational activities (so-called "recreational" services). These ES were defined as the biophysical characteristics of agricultural ecosystems contributing to the creation or maintenance of settings suitable for the pursuit of recreational activities.

ES linked to the pursuit of recreational activities without taking of fauna or flora were differentiated from those linked to the pursuit of recreational activities with taking (hunting, gathering of wild animals or plants), since the biophysical characteristics and benefits obtained from these two ES are different in nature.

The analysis of "recreational services with taking" has not been developed here, due to a lack of data allowing for a detailed characterisation of ecosystems' capacity to serve as habitat for the relevant wild faunal and floral species. Moreover, the link between wild fauna and agricultural ecosystems is difficult to establish since wild animals can be highly mobile across ecosystem types (agricultural ecosystems, wetlands, forests, etc.).

▌ Characteristics of agricultural ecosystems as opportunities for recreational activities without taking

Recreational activities "without taking" include several types of outdoor pursuits: *in situ* wildlife or nature study, walking/hiking, agritourism, etc. These activities can be pursued in spaces managed specifically by humans for recreational purposes (such as urban parks) but can also be undertaking in landscapes composed of a mosaic of ecosystems. It is therefore a question of characterising and evaluating the recreational potential of agricultural ecosystems.

The relationship between outdoor recreational activities and ecosystems is strongly linked to the idea of landscape, defined by the European Landscape Convention as "a part of the land as perceived by local people or visitors". Landscape here refers to an ensemble of material, biophysical attributes, but it is observers' perceptions of this ensemble – how they look at it and the cultural associations they attribute to it – that determine whether a landscape is considered to be pleasant or unpleasant. The public establishes a link between biophysical elements and the recreational opportunities they provide based on a positive perception of or a preference for this material reality.

The pursuit of outdoor recreational activities in most cases requires that individuals travel from their place of residence to the ecosystems that offer these services. As a result, the exploitation of ecosystems' recreational potential, and thus the level of ES effectively supplied by these ecosystems, will also be affected by their accessibility to the public.

Biophysical determinants and external factors

Biophysical determinants

Agricultural ecosystems create specific landscapes (woods and fields, open grasslands, viticultural landscapes, etc.) composed of biophysical elements: cultivated areas, areas of natural and semi-natural vegetation (hedges, field borders, ditches, etc.), rocky or barren areas. The spatio-temporal distribution of crops, together with field layouts and the structure of semi-natural habitats (slope, hedges, wooded areas, isolated trees, etc.), determine the landscape's structure, and are the key biophysical determinants of the landscape's capacity to supply the ES for recreational potential.

Natural heritage features and built elements located in the vicinity of agricultural landscapes (buildings, barns, stone walls, etc.) are likewise key landscape components as perceived by the public. It is often difficult to distinguish between the roles of ecosystem elements vs. built elements within the attractiveness of agricultural ecosystems for recreational pursuits.

External factors

The principal external factors in the level of supply of this ES are the human activities that determine the structure of the landscape. With regard specifically to agricultural activities, practices relating to the management of exported biomass (harvest, management of crop residues, etc.) modify the appearance of ecosystems and therefore their recreational potential.

It should be highlighted that infrastructure and facilities (cycle paths, hiking trails, animal observation sites, etc.) are necessary to access the sites and benefit from their recreational potential.

Level of ES provision

Within the framework of the MAES program, the assessment of this ES is based on socio-economic indicators of recreational use: the number of visitors to agricultural spaces, the number of rural businesses offering services to tourists, the number of kilometers of pedestrian walkways or cycle routes. Such variables allow one to estimate the level of ES effectively supplied by landscapes overall, but not to quantify the specific contribution of agricultural ecosystems.

Work conducted by the European Joint Research Center (JRC) proposes an indicator combining i) an estimate of the site's recreational potential based primarily on the idea of the "naturalness" of the environment, and ii) an estimate of the site's proximity to residential areas. This indicator is then analysed with respect to an indicator of potential demand, represented as the number of potential trips made by individuals living in proximity to the site.

Assessment method

Using the methodology developed by the JRC (Paracchini *et al.*, 2014),[43] the potential level of ES supplied by agricultural ecosystems was assessed by calculating their degree of naturalness at a detailed spatial scale (national territory subdivided into 100 m pixels). The "naturalness" indicator estimates the level of human intervention relative to the potential "natural" ecosystem.

The recreational potential of agricultural ecosystems

Figure 4-6 shows a cartographic presentation of the degree of naturalness of French agricultural ecosystems as an indicator of their capacity to provide a setting for the pursuit of recreational activities.

Less than 5% of the UAL is associated with a high degree of naturalness (classes 2 and 3). These ecosystems correspond to natural pastures and meadows located in high mountain areas (Alps/Pyrenees) and in some parts of the Massif Central (Grands Causses, the southeastern part of Limousin). The great majority of agricultural ecosystems in France show a medium degree of naturalness (vineyards and grasslands in other parts of the Massif Central, Lower Normandy, and the eastern mountains) or a low level of naturalness (major field crop areas in western and southwestern France).

Figure 4-6. Degree of "naturalness" of agricultural ecosystems

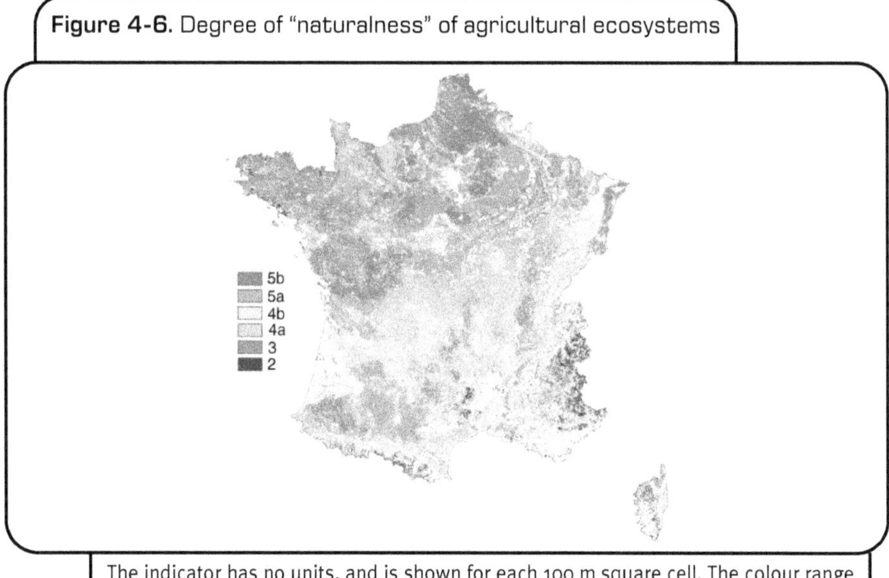

The indicator has no units, and is shown for each 100 m square cell. The colour range varies from red (low degree of naturalness) to green (high degree of naturalness).

---

43. Paracchini M.L, Zulian G., Kopperoinen L., Maes J., Schägner JP, Termansen M., Zandersen M., Perez-Soba M., Scholefield P.A., Bidoglio G., 2014. Mapping cultural ecosystem services: A framework to assess the potential for outdoor recreation across the EU, Ecological Indicators 45:371-395.

This indicator assumes a positive correlation between ecosystem attractiveness and ecosystem naturalness. Some types of agricultural ecosystems, however, such as land areas used for grasslands, viticulture, or tree fruit production, are characterised by a moderate or low level of naturalness but have landscape attributes recognised as being appealing for recreational pursuits (for example, grassland areas with scattered trees as found in Lower Normandy, or the vineyard areas in the Loire Valley or in Burgundy). The level of "naturalness" is thus insufficient as a way of assessing the recreational potential of agricultural ecosystems, and needs to be complemented with a descriptor of the cultural dimension associated with certain types of agricultural ecosystems or agricultural landscapes (mosaics of ecosystems dominated by agricultural ecosystem types).

Towards a quantification of the effective level of ES supply

Regardless of ecosystem type, exploitation of an ecosystem's recreational potential depends on the ecosystem's accessibility. The level of ES effectively supplied to society can thus only be quantified by also taking into account the accessibility of these agricultural ecosystems.

In the absence of representative data on the frequency of recreational visits to agricultural ecosystem locations at the level of France as a whole, the study made use of an indicator proposed by the JRC for the potential frequency of visits to such sites. This indicator may be calculated for any country. It consists of estimating the potential number of short day trips (of approximately 8 km at most) between residential areas and areas offering the opportunity to pursue recreational activities. It is based on the assumption that the likelihood of a trip from one's home to another location falls as the distance (measured as the crow flies) between the two increases. This indicator could thus be used to calculate the number of potential trips from residential areas ("urban" pixels) to agricultural ecosystems or landscapes ("agricultural" pixels). Weighting the recreational potential of agricultural ecosystems according to their potential visit frequency estimated in this way thus provides an initial idea of the extent to which people are likely to benefit from the recreational potential of agricultural ecosystems located near their place of residence.

In addition to developing a more robust indicator of the recreational potential of agricultural ecosystems than their degree of naturalness, a number of improvements could be made to the indicator for the potential frequency of visits to these ecosystems.

- Calculating distances in terms of travel time rather than as the crow flies. It seems fair to assume that individuals are only inclined to travel for the pursuit of recreational activities without taking if the trip is below a certain travel time, and that the likelihood of making a trip from one point to another decreases as travel time increases.

- Taking into account the willingness of individuals to make a given journey depending on whether recreational activities are pursued in the context of daytrips or as longer excursions. This consent to move may partially depend on the demographics of the urban residential areas (notably population density and socio-professional class).

- Taking into account the "internal" accessibility of the sites through the trails and hiking trails located within their area, and which could be characterised using the TOPO® database.

5. The economic assessment of ecosystem services: Cautions and challenges

In EFESE-AE, the economic evaluation adopts the approach developed in the scientific literature consisting of assigning a value to ecosystem services based on the biophysical quantification of their level of provision.

This makes it possible to develop a rough idea of the costs that would be incurred if recourse to substitution technologies became necessary (e.g., inputs of synthetic nitrogenous fertilisers to substitute for ES for the supply of mineral N to crop plants), or the value of damages that would result from the disappearance of the ES (e.g., yield losses in the absence of ES for the regulation of pest insects).

An analysis of economic assessment methods for ES was carried out for eight of the twelve regulating ES examined in EFESE-AE, giving priority to those for which a biophysical assessment had been completed. The results obtained are preliminary and should be used with caution. They aim above all to illustrate the methodological as well as conceptual difficulties associated with the economic evaluation of ES.

 **Available methods**

Only those ES that are currently benefiting society or farmers were assessed in EFESE-AE. The so-called "Revealed preferences" methodologies,[44] which are based on the observation of the actual market behaviors of ES beneficiaries, and thus on information that is generally objective and readily available, were favored. For ES that are not currently being exploited but which are likely to be exploited in the future (genetic resources of soil microbiota, etc.), economic assessments should be based on explicit scenarios involving key assumptions with regard to future policy decisions or other changes at the national or international level.

A common approach in economics is to assume a degree of correspondence between observed market behaviors (that is, the quantities of goods offered and demanded for exchange) and the satisfaction each individual receives from that exchange, measured in terms of utility for the consumer and profit for the producer. When a good or a service[45] is directly exchanged on a market, it is typical to use its market price as an

---

44. Methods that allow one to obtain values *ex post*, estimated indirectly based on prices obtained in the relevant markets for goods whose consumption is linked to the ES in question.
45. In the sense of national accounting.

indicator of its economic value. The unit price corresponds to the meeting point between the quantity offered for sale and the quantity demanded for purchase. The economic value of agricultural goods exchanged on markets is thus classically measured through their market price.

In contrast, ES do not become the object of market transactions. The demand from society (or from farmers, depending on the benefit in question) is thus not directly observable. It is possible, however, to observe variations in the consumption of certain market goods that are linked to these ES, such as the external inputs that can be substituted for input ES. Estimating the value of input ES based on the demand for external inputs is central to the development of approaches based on the use of production functions, which seek to characterise the benefits received from ES in terms of variations in profit (due to savings made on input purchases and the variation in productive output). Application of this approach thus depends on elaborating the chains of cause and effect existing between variations in ES, the behavior of producer agents, the use of substitution technologies, and variations in agricultural production. The use of such methods for economic assessment will thus only be possible when our understanding of the interactions among ES, socioeconomic behaviors, and agricultural production is sufficiently advanced.

In situations where information on these interactions is lacking, two methods are frequently made use of to develop economic assessments of ES.[46]

The first is to estimate the replacement cost of the ES in terms of an alternative technology that can be used to compensate for its absence (or disappearance). The replacement cost is thus determined by the market price of the technology used as an alternative to the ES. This approach was employed to assess the ES for the supply of mineral N and for the storage and return of water to crop plants. In these two cases, biophysical assessment makes it possible to estimate the quantity of inputs (respectively, nitrogenous fertilisers and irrigation water) these ES allow the farmer to not use. The replacement cost method can provide a reliable estimate of the economic value of the ES, assuming that the alternative technology selected for the assessment supplies the same benefit as the ES, is reliable, is socially acceptable, and costs less than or the same as the value of the benefit supplied by the ES. If one or more of the conditions are not met, the most logical option for the farmer would be to reconfigure the agricultural ecosystem (and thus discontinue or modify the production system) rather than maintaining production "at all costs" in the absence of the ES.

---

46. Two other methods associated with revealed preferences methodologies are poorly suited to the assessment of regulating ES. Transportation costs methods are only applicable to the assessment of landscape recreational potential, through the amounts of money spent to access a given location and to engage in a recreational activity there. Hedonic pricing methods are used to estimate the contribution of the environment to the value of a good, examining the effect that the environment has on the price of this good. Hedonic pricing methods are mainly applied to the price of real estate in environments with specific characteristics (landscape attributes, occasional pollution, etc.). Beyond their relevance to specific types of economic objects, these methods are difficult to apply at a large scale.

The other method estimates the cost of damages avoided as a result of the presence of the ES. As applied to input ES, this involves estimating the cost of the production losses that would occur if the agricultural ecosystem did not supply (or no longer supplied) the ES, assuming that no steps would be taken by the farmer to compensate for the ES's disappearance. Another, equivalent way of approaching the problem is to quantify the percentage of agricultural output attributable to the ES (all other things being equal). The economic value is then obtained through the market price of the agricultural goods in question. This method was used to assess the percentage of agricultural production attributable to the ES for crop pollination, and, in an exploratory fashion, the percentage attributable to the two input ES "N and water," taken together.

A limitation of approaches making use of market prices lies in the fact that they assume that such prices are reliable indicators of the social demand for and rarity of the goods and services used as replacements for the ES. In practice, market prices are frequently also affected by a variety of social and political factors, including subsidies, and are thus not necessarily an accurate reflection of the preferences of consumers and/or citizen.

Finally, because of the diversity of assessment methods employed, it is not possible to sum the economic values calculated for each ES as a way of estimating the overall value of ES provided by agricultural ecosystems.

The economic assessment of input ecosystem services' contribution to agricultural production

In EFESE-AE, an economic value was calculated for three input ES: i) the two ES for the supply of mineral N and the storage and return of water to crop plants, and ii) the ES for crop pollination where the pollination contribution to yield output is measureable.

▌Proposed methodologies for assessing the supply of mineral N and storage and return of water to crop plants

The replacement costs method was used to assess ES for the supply of mineral N and the storage and return of water to crop plants. With regard to N, the optimal response of the farmer to a deficit in the ES for the supply of mineral N (assuming that the initial level of supply was optimal) is to compensate for the deficit through the application of external fertiliser inputs.[47] The same logic may be applied to the ES for the storage and return of water to crop plants.

---

47. In economic terms, this is to assume that the farmer is rational in the sense that he or she will apply the amount of N that will provide maximum profit, and that the marginal yield return on N inputs decreases as the amount applied increases (each additional unit of N will result in a smaller yield gain than the preceding unit of N).

We can note that as calculated with the avoided damages method, the economic value of these ES would be the value of the percentage of production that would be lost in the absence of the ES. However, the simulation scenario developed in EFESE-AE did not allow for an estimate of production levels in the absence of each of these two ES. It did make it possible, however, to quantify the amount of agricultural production attributable to these two ES taken together (all other things being equal – see Chapter 3). This quantity of production may thus be considered as the "damages avoided" due to the presence of these two ES. The value of this quantity of production was then estimated, and may be used as a point of comparison with the economic assessment of the two input ES taken individually with the replacement costs method (Box 5-1).

The biophysical assessment of the ES for the supply of mineral N and storage and return of water to crop plants allowed for the quantification of the levels of these two ES: the average annual quantity of mineral N supplied by the ecosystem during the period of crop growth (through symbiotic fixation and mineralisation), and the average annual quantity of water transpired by the commercial crop. An economic assessment of these ES can thus be provided by estimating the cost of (i) the synthetic nitrogenous fertiliser and (ii) the irrigation water it would be necessary to supply to maintain the same level of production in the absence of these ES, assuming that the agricultural ecosystem manager compensates for the absence of these ES with optimised inputs relative to crop needs. The replacement costs for each of these ES, for a given crop and geographical location, may be calculated as follows:

- for the ES for supply of mineral N to crop plants:

Economic value of the ES ($EVES_N$ in €/yr/ha) = Average quantity of N supplied by the ecosystem to the crop (in kg N/ha/yr[48] x Market price of N (€/kgN)

- for the ES for storage and return of water to crop plants:

Economic value of the ES ($EVES_{water}$ in €/yr/ha) = Average quantity of water transpired by the crop (in m³/ha/yr) x Irrigation cost (in €/m³)

Agricultural statistics provide ready access to an average market price for N (€0.85/kgN for the period from January 2008 - January 2016), but the cost of irrigation is more difficult to establish. Irrigation costs are highly variable depending on both the water source and the type of irrigation equipment. An analysis of irrigation costs in France as reported in the gray literature suggests that average irrigation costs in France vary from €0.04/m³ to €0.335/m³, with no indication as to how costs vary in different parts of the country. Clearly, however, irrigation costs do vary substantially by geographic area. In the absence of more precise data, these two values were used as a range.

The results of the economic assessment of the two ES are presented in Table 5-1. These results should be understood as an order of magnitude of the replacement cost for these two ES for agricultural ecosystems dedicated to the eight major crops grown in

---

48. The biophysical assessment provided an annual average quantity of N supplied and water returned by the ecosystem at the level of the cropping system as a whole, not crop by crop. An initial calculation is thus required to apply a change of scale procedure to these results, so as to obtain values per crop and per PCU.

France (representing 91% of land area dedicated to major field crops and industrial crops within mainland France).

Table 5-1. Average annual values for the ES "supply of mineral N" and "storage and return of water to crop plants" for France as a whole, estimated for eight crops using the replacement costs method

ES: / Crop:	ES supply of mineral N to crop plants (€/yr/ha)	ES storage and return of water to crop plants (€/yr/ha)		Total crop area = land area occupied by the crop for France as a whole (in ha; average LPIS 2010-12)
		Minimum cost	Maximum cost	
Sugar beet	103	7	59	437,165
Soft wheat	61	6	46	6,006,826
Barley	63	6	46	1,548,366
Oilseed rape	75	7	58	1,590,907
Forage maize	83	7	55	1,264,859
Grain maize	116	7	62	1,643,784
Spring peas	149	5	42	291,370
Sunflower	59	4	29	722,950

Values listed are averages for the period 2010-2012. For each crop, the representativeness of the land area used for the simulation relative to the total crop area for France as a whole was calculated by dividing the land area occupied by the crop in the PCUs where it was simulated by the total land area occupied by the crop (as obtained from the French LPIS).

Again, these values should not be used to make extrapolations at the national level. Calculating a national replacement cost value for ES involves imagining extreme situations in which it would necessary to fertilise and/or irrigate all crops in response to the total disappearance of these two ES. In addition to the improbability of such a situation in biophysical terms, the total absence of one or both of these ES would necessarily have major impacts on the availability and price of synthetic fertilisers and of water for irrigation, with knock-on effects on farmer behavior, neither of which are accounted for in the calculations made at the per hectare level. By way of comparison, the cost of extreme drought across all of France would in all likelihood be a total crop loss. It therefore makes sense to limit the replacement cost for irrigation and fertilisation to their opportunity cost – that is, the difference between profit margins realised by farms in the current situation (in the presence of the ES) and profit margins realised in an "absence of ES" situation. In other words, beyond a certain threshold, it would be more rational in economic terms to change crops or to plant no crop; i.e., to reconfigure the agricultural ecosystem.

Box 5-1. Economic value of the crop production attributable to the input ES "N and water"

The economic value of the part of crop production made possible by the two input ES "N and water" considering together (EV_{ESprod}) was calculated as follows for each of the crops covered by the analysis:

$$EV_{ESprod} \text{ (€/y)} = \text{Average percentage of production attributable}$$
$$\text{to the input ES "N and water"}$$
$$\times \text{ Average yield for the years 2010-2012 (t/yr)}$$
$$\times \text{ Average land area for the years 2010-2012 (ha)}$$
$$\times \text{ Average price for the years 2010-2012 (€/t)}[49]$$

As a reminder (see Chapter 3), the level of crop production made possible by the input ES "N and water" corresponds to average annual yields estimated with the STICS model in the absence of all inputs (no irrigation, nitrogenous fertilisation, soil tillage or incorporation of crop residues). For the purposes of biophysical quantification, the level of production attributable to the two input ES "N and water" is expressed relative to total production (i.e., total production attributable to the combined effects of these two ES + external inputs of water and N). For the economic assessment, however, it is the absolute level of production attributable to these two ES that is considered as the "damage avoided" due to the presence of these ES.

The value of EV_{ESprod} obtained for each crop was extrapolated to the total land area for that crop for France as a whole. The results of this assessment are presented in Table 5-2.

The seven crops listed in Table 5-2 account for 89% of land area in major field crops and industrial crops in mainland France for the period 2010-2012. Taking all crops together at the national level, the annual average value of the part of production attributable to the input ES "N and water" is on the order of €9.8 billion, or 50% of the average total value of production for the corresponding land area (total value of production = €19.6 billion).

The reliability of the values obtained for each crop depends on the robustness of the STICS simulations and on the accuracy of the extrapolation. The latter depends on the representativeness of the simulated land areas relative to the total land area for each crop. For some crops, however – especially grain maize, sunflower and forage maize – the cultivated land area taken into account in the simulation is low compared to the actual total land area planted to these crops in France. This is because these crops are only weakly represented (in terms of land area) within the dominant crop rotations that were simulated with the STICS model for each PCU (see Box 2-1). Simulation of a larger number of rotations in a larger number of PCU would improve the representativeness of the biophysical estimates and thus of the economic assessment. The additional simulations could be organised so as to better include land area in crops poorly represented in the current simulation plan.

Due to the differences between the assessment methods used to assess the two input ES "N and water" individually, on the one hand, and to assess the economic value of crop production attributable to these two ES together, on the other hand, the two sets of results are not directly comparable. In addition to the limitations already mentioned, the results obtained for the two ES "supply of mineral N" and "storage and return of

---

49. Estimated using the database FAOstat.

water to crop plants" evaluated separately with the replacement costs method cannot be compared since their assessment is not based on the same reference situation nor on the same type of substitution technology. Finally, to sum these two values would amount to assuming that the production factors at stake are completely substitutable, which is not the case (due to interactions between the biophysical processes relating to N and water). In other words, a direct comparison of the sum of the two first values with the third value is not meaningful. The two sets of results provide complementary information relating to the assessment of the two input ES for water and N.

Table 5-2. Average annual values for the production attributable to the input ES "N and water" estimated for seven crops at national level (France)

Crop:	Average part of production attributable to the input ES "N and water" (%)	Average value of production attributable to the input ES "N and water" for mainland France (M€/yr)	Average total value of agricultural production for mainland France (M€/yr)	Representativeness of the land area used for the simulation
Sugar beet	34	456	1,373	69%
Wheat[a]	58	4,917	8,605	66%
Barley[b]	58	1,027	1,794	72%
Oilseed rape	28	615	2,320	68%
Forage maize[c]	67	1,093	1,589	31%
Grain maize	41	1,173	3,129	53%
Sunflower	68	539	757	53%
TOTAL	/	9,822	19,567	

For each crop, the representativeness of the land area used for the simulation relative to the total crop area for France as a whole was calculated by dividing the land area occupied by the crop in the PCUs where it was simulated by the total land area occupied by the crop (as obtained from the LPIS). Data on the average price of peas were not available from the FAO database, so this crop was excluded here. Peas accounted on average for about 2% of land area in major field crops and industrial crops in France for the period 2010-2012.
a. Land area in hard wheat was assimilated to land area in soft wheat when calculating the part of production attributable to the ES and when calculating value of agricultural production.
b. Land area in barley was assimilated to land area in soft wheat when calculating the part of production attributable to the ES.
c. The economic value of forage maize was estimated relative to grain maize by using a coefficient of 0.5 to convert average yields in metric tons of forage maize dry matter per hectare (t DM/ha) into the equivalent tons of grain maize per hectare (t/ha), and then applying the price of grain maize (€/t).

Furthermore, in contrast to the ES for the storage and return of water to crop plants, the biophysical assessment of the ES for supply of mineral N to crop plants only allowed for quantification of the potential level of ES supplied by the agricultural ecosystem, and not the level of ES actually used by the farmer. Calculated on this basis, the assessed economic value is thus overestimated. In addition, this approach assumes that the farmer will compensate for the absence of the ES "supply of mineral N to crop plants" exclusively through the addition of synthetic fertilisers. In reality, however, other fertilisation strategies could also be made use of, such as the application of organic fertilisers or the reduction of biomass exports. The magnitude of the replacement costs could also lead the farmer to reconfigure the agricultural ecosystem, for example by introducing a legume as a commercial crop or by planting a cover crop. The elaboration of potential adaptation strategies combining these three levers followed by an assessment of such strategies' biophysical and economic effects was beyond the scope of EFESE-AE, however.

▌ Updating the economic assessment of crop pollination

Because of the importance of pollination in the production of a large number of different crops, especially fruits and vegetables, several proposals for the economic assessment of this ES may be found in the literature. In EFESE-AE, the analysis focused on the pollination of crop species. For the purposes of the economic assessment, only the benefit derived from this ES by the farmer was considered.

An assessment of this ES using the replacement costs method would require identifying available alternative technologies and their corresponding implementation costs for each crop. Currently available alternatives (e.g., manual pollination) have implementation costs that are higher than their opportunity cost, implying that farmers would discontinue production rather than seeking to compensate for the absence of this ES. For this reason, the avoided damages method was used to estimate the economic value for this ES.

Assessment method

Some studies have sought to extrapolate the yield impacts of pollination deficits based on experiments or contexts in which pollinators are excluded. Although this approach appears promising, the results are too partial to allow for standardisation. The method used here, is based on the use of the ratios of crop yield dependence on pollinators estimated under controlled conditions as proposed by Gallai *et al.* (2009).[50] These have also been adopted by the Food and Agriculture Organisation (FAO). This method is thus based on the same data used to calculate the pollination index in order to estimate the effective level of this ES (see Chapter 2). Eighteen pollinator-dependent crops (or crop categories), yields for which were available (mean values at the departmental level in Annual Agricultural Statistics), were considered for this analysis. They include fruits, vegetables, and oil crops, and amount to a bit less than half of the total number of crops for which ratios of pollinator dependence have been

---

50. Gallai N., Salles J.M., Settele J., Vaissiere B.E., 2009, Economic Valuation of the Vulnerability of World Agriculture Confronted with Pollinator Decline, *Ecological Economics*, 68(3) : 810-821.

established. The remaining crops could not be considered because they do not appear in the FAOSTAT database used to assemble price data. Seed crop production was also excluded due to a lack of data. These two exclusions from the calculation thus lead to an underestimation of the overall economic value of insect pollination services (EVIPS).

Results and cautions for interpretation

The aggregate value of the ES for crop pollination averages €2 billion per year for the years 2010-2012 (with a slight variation across the three years). Figure 5-1 shows the spatial distribution of the average EVIPS for these three years at the departmental level.

The variation in the value of the ES across departments may be attributed in large part to two factors:

• the difference in the total value of agricultural production for each department, which is affected both by differences in agricultural land and by prevailing crops/ management methods on farms;
• the relative importance of pollination-dependent crops within the total value of agricultural production for each department.

Figure 5-1. Spatial distribution of the average annual economic value of the ES for crop pollination, for the years 2010-12

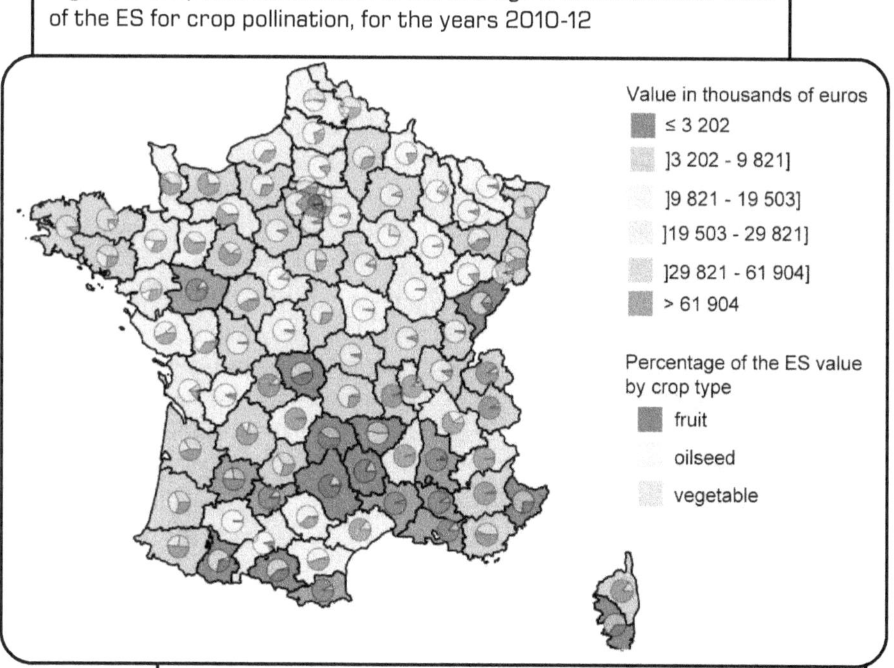

EVIPS values (and percentage by major crop type) for the three departments in the Île-de-France: Department 92 = €8,136 (100% fruit crops); Department 93 = €38,683 (100% oilseed crops); Department 94 = €237,236 (75% fruit crops, 17% oilseed crops, 8% vegetable crops).

Thus, departments in which the value of the ES appears to be the highest are those where fruit production and oil crop production, and to a lesser extent vegetable production (especially cauliflower), are most prevalent.

The method used here appears to be relatively simple and seems to require little input data, but the indicator is associated with certain biases that should be taken into consideration when looking at the results. An implicit assumption is made that the ES for crop pollination is supplied at its maximal level for all areas and all crops. However, results from the biophysical assessment tend to show that pollination can be a significant limiting factor for the production of agricultural goods (see Chapter 2). A hypothetical calculation can be used to illustrate the effect of this bias on the results of the economic assessment (Figure 5-2). Imagine a crop for which the ratio of yield dependence on insect pollination is 40%, and for which "maximum" production with no pollination deficit (all other things being equal) is 100 units. If there is a pollination deficit, and the level of ES effectively supplied to the farmer is only half of its maximum, the harvest will be reduced by half of 40%, and the observed production (as reflected in Annual Agricultural Statistics) will only be 80 units. Assuming that the observed production corresponds to the "maximum" production, however, will lead to an EVIPS calculation that overestimates the value of production attributable to the ES. In other words, when the ratio of pollination dependence is applied to a level of production not limited by pollination, it will produce a correct estimate of the economic value of the ES. If, on the other hand, it is applied to a level of production that is limited by pollination, it will tend to result in a biased estimate situated between the effective ES and the maximum ES. The higher the pollinator-dependence ratio and the higher the pollination deficit, the greater the overestimate of the economic value of the ES based on the EVIPS indicator will be. Subject to having a generic and reliable indicator of the pollination deficit, this information could easily be integrated into the calculation of a EVIPS which would then be corrected for the bias identified above.

Calculation of the EVIPS is highly contingent on the quality of the crop's pollinator-dependence factor, which is only known in an indirect way. Estimates of the effects of pollination on yields are also assumed to occur "all other things being equal" – a standard assumption in economic analyses, but one that fails to account for any interactions that may exist between the ES for crop pollination and other factors, such as variety choices or other agricultural practices.

Finally, one should keep in mind that pollination services also relate to wild flora and thus are indirectly linked to a range of other ES (e.g., cultural ES) that benefit society as a whole but are not taken into consideration in the EVIPS indicator.

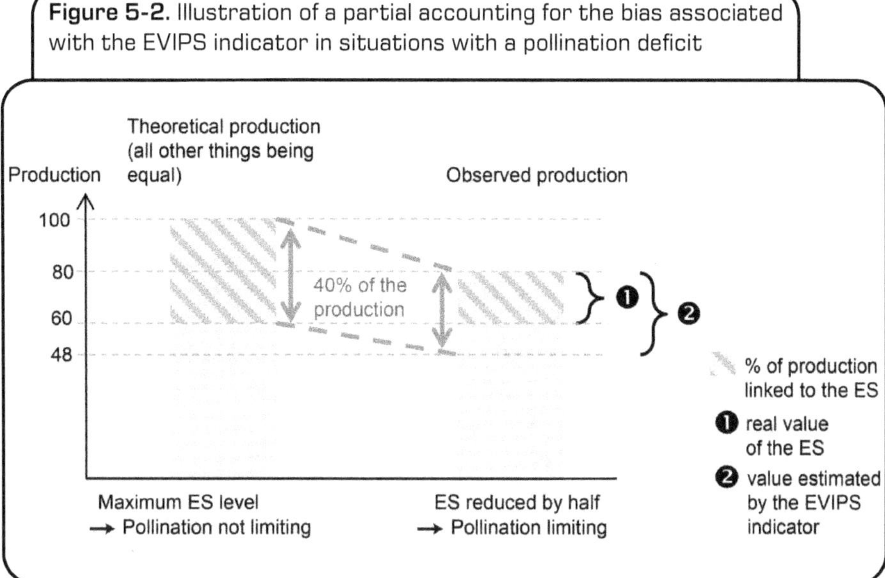

Figure 5-2. Illustration of a partial accounting for the bias associated with the EVIPS indicator in situations with a pollination deficit

The economic assessment of an ecosystem service provided to society: the example of global climate regulation

It is recognised that ES for global climate regulation cannot be assessed with a replacement costs method, given the uncertainties surrounding the cost, reliability, and social acceptability of C storage technologies as well as the wide range of different contexts to be considered (pedoclimatic contexts, production systems, etc.). Economic assessments of ES for global climate regulation are thus more typically approached in terms of the reduction or avoidance of GHG emissions, assigning a value to each ton of C (or of CO_2) not emitted into the atmosphere. Given the temporal gap between the implementation of measures to reduce GHG emissions and observable impacts in terms of the quantity of GHG present in the atmosphere (and resulting impacts on the climate), estimates of the economic value of this ES are based on the construction of scenarios for various changes in the socio-economic context (demographics, land use, etc.). In this approach, the assessment strategy varies according to the definition given to the "value": the cost associated with a reduction in GHG emissions, the market price of C (which assigns a price per ton of non-emitted C for sectors covered by these markets), the social cost of the C (the marginal cost of economic damages resulting from higher concentrations of GHG in the atmosphere), or again the cost of actions taken to avoid emission of one ton of C or to sequester one ton of C. Considering the range of methods for "costing" C linked to each of these "values," the "State-imposed Value of Carbon", as estimated by the Quinet Review for the purposes of establishing the level of the C tax in France (levied on emitters of CO_2 based on the polluter-pays principle), seems to offer the best compromise. It represents the value used for the majority of public investment purposes in France.

On the biophysical level, two key components of the ES for global climate regulation were defined: i) a service for the conservation of stored C; and ii) a service for the storage of additional C. Further work remains to be done to pair the two corresponding indicators. Following the same logic, two economic values could be considered: i) the value of the total current stock of C in the ecosystem, and ii) the value of the annual flow of C stored or emitted by the agricultural ecosystem. This distinction is important insofar as the quantities of C involved are radically different. In addition, it raises questions about the permanence of C storage: a quantity of C stored per year t by the agricultural ecosystem may or may not be stored indefinitely, whereas the price of C is generally applied per ton of C never to be emitted. Assessing annual flows of sequestered C makes it possible to set this question aside: what is emitted by the system is considered as a flow with a negative effect on the climate.

The decision to assess the stock vs. the flow of C within an ecosystem may also be determined by the existence of potential threats or mutation weighing on the latter: if an ecosystem is in danger of disappearance, for example because of a radical change in land use, seeking to assess the total quantity of stored C in the ecosystem is a step toward evaluating the amount of sequestered C at risk of being emitted into the atmosphere. If the objective is to evaluate changes in agricultural practices or more marginal threats to the ecosystem, an analysis in terms of C sequestration over a given period of time is probably more relevant. It may be desirable to evaluate the two components of the ES for global climate regulation noted above: first, the function of annual C sequestration, allowing for a reduction in the quantity of GHG present in the atmosphere; and second, the function of protection offered by C storage over the long term.

A calculation of the economic value of C flows associated with agricultural ecosystems (V_{flow}) for time horizon T could take the following form:

$$V_{flow} = \sum_{t=0}^{T} \frac{EV_t \times NCS_t}{(1+a)^t}$$

where EV is the economic value of a ton of CO_2 considered for the year t (in constant euros), *NCS* is the net quantity of C stored in the agricultural ecosystem at year t, and *a* is the social discount rate.[51]

In cases where C stocks are to be evaluated, it is necessary to take into account the temporary nature of C storage in the ecosystem. One solution would be to calculate the percentage of the total quantity of stored C *not* threatened by a change in practices

---

51. In the literature, this rate is generally set at 4 or 4.5%.

(or in ecosystem configuration), and then to assign the remainder no value since it will not be stored for the long term. In other words, the idea is to only value C stored for the very long term (at least 30 years). A calculation of the economic value of long-term C storage (*Vstock*) over time horizon *T* could take the following form:

$$V_{stock} = \sum_{t=0}^{T} RR \times \frac{EV_t \times CS_t}{(1+a)^t}$$

where *TR* represents the rate of return for this C capital stored over the very long term[52], EV the economic value of a ton of CO_2 considered for the year *t* (in constant euros), CS the part of the C stock considered to be immobilised over the very long term by the ecosystem, and *a* is the social discount rate. One study applied to forest ecosystems, conducted by the French Center for Strategic Analysis, estimated the percentage of C stored for the very long term in woody biomass at approximately 25% and the percentage stored for the very long term in the soil at approximately 75%. In the case of major field crop and grassland agricultural ecosystems, accounting for the amount of C stored for the very long term aboveground seems, generally speaking, not that relevant. The percentage of C stored for the very long term in the soil remains to be determined.

Economic assessments are challenging due to lack of matching between biophysical and economic indicators

In contrast to the ES presented previously, in situations where the biophysical quantification of an ES is based on an indicator of the condition of one of its biophysical determinants (e.g., beetle abundance as an indicator of the weed seed regulation level), rather than on a direct indicator of the level of ES supply, the avoided damages and replacement costs methods cannot be applied. Four ES fall into this category: the two ES for biological control by conservation supplied to the farmer, the ES for the regulation of drained water quality supplied to society, and the ES for soil stabilisation and erosion control, supplied to both categories of beneficiaries.

∎ Biological controls by conservation

First, we should note that all of the work assembled on economic assessment methods for ES of biological control relate to ES for the regulation of insect pests, and none to the regulation of weed seeds. The avoided damages method is typically the approach adopted in these studies. Methods considering the cost of agronomic

---

52. A rate of 4% is suggested in the literature.

practices (pesticides, mechanical cultivation, etc.) to substitute for ES (replacement costs methods) are also used.

In EFESE-AE, the indicators used to quantify ES for biological control make it possible to predict the density of pest species (number of weed seeds, rate of aphid population growth) or of their predators (beetles), but do not allow for an estimate of the damages caused to crops by pests. The avoided damages method can thus not be used here. Linking these biophysical indicators to associated yield losses would require the use of damages functions, which are generally unavailable and difficult to establish for the large number of relevant crop/pest pairs. National and international research efforts are currently focused on quantifying these relationships for the most important pests of major field crops. Assuming that these damage functions can be established, economic assessment with the avoided damages approach leads to an assumption that farmers do not alter their practices based on ES levels, since the assessment would seek to estimate the level of pest damage in the total absence of the ES, but with the same agricultural practices.

Furthermore, in contrast to the ES for the supply of mineral N and the storage and return of water to crop plants, the relationship between biological regulation and pesticides is complex, making it difficult to establish a firm link between the ES level and the alternative technologies that could be substituted for it. That is, the quantity of pesticides that could compensate for the absence of the ES cannot be inferred from the indicators for the ES developed in EFESE-AE. It is thus impossible to use the replacement costs method. Another issue is that pesticides substitute for ES by reducing pest damage, but they are also likely to have negative effects on the structure or the activity of auxiliary species communities (pest predators and parasitoids). ES levels also depend on the wider landscape in which the field is situated and thus on the overall level of agricultural intensification, which is rarely included in the assessments.

One possible option for future economic assessments of ES for biological control would be to define how agricultural practices are altered in response to variations in ES levels. An integrated model – connecting economic, agronomic and ecological approaches – would allow for the simulation of missing data and the assessment of ES according to a number of clearly identified assumptions. Ideally, such a model would be calibrated to take into account local specificities of ES for biological regulation. Key entry variables for such a model would include:

- parameters representing the lifecycle characteristics of pest populations and pest predators (growth rates, predation rates, mobility, etc.), so as to be able to simulate population dynamics according to different land use and agricultural practices scenarios;
- parameters for calibration according to different damages functions, to simulate the effects of pests on crops;
- parameters for an economic model of land use and agronomic practices.

Such a model could be used according to different economic decision scenarios, and a sensitivity analysis for the principal parameters would make it possible to arrive at ranges of probable values for ES at different geographic levels. Finally, such a model could be improved over time as research advances on the economic and ecological phenomena that underlie different ES.

▮ The regulation of drained water quality

The quality of aquatic environments has been the focus of a large number of studies seeking to estimate their value, although these have focused primarily on wetlands, whose characteristics are far removed from those of agricultural ecosystems. Another category of research has sought to put a price tag on the environmental impacts of agricultural practices, rather than seeking to value the ES provided by these ecosystems. This is the case, for example, with research conducted by the French Ministry of Ecology on the costs of the most significant forms of agricultural pollution. This work has sought to evaluate the costs incurred from the contamination of ground and surface waters by agricultural pesticides and excess N, leading to a loss of quality of aquatic environments and water resources and necessitating more expensive treatment procedures for drinking water.

A study led by the French Center for Strategic Analysis estimates the economic value of agricultural ecosystems' capacity to supply healthy vs. polluted water using the replacement costs method, and drawing on three types of data: i) average annual consumption of potable water; ii) the cost of avoided water treatments; iii) average ecosystem contribution to water quality, corresponding in this approach to the quantity of water purified by the ecosystem. However, the biophysical indicator adopted in EFESE-AE for the ES "water quality regulation with respect to N " does not correspond to a quantity of good quality water restored by the agricultural ecosystem, but to a quantity of non-leached N. It therefore remains to develop methods capable of using this biophysical indicator as a starting point for economic assessment.

▮ Soil stabilisation and erosion control

Economic assessment of ES provided to the farmer

Using the avoided damages method to provide an economic assessment of the ES for soil stabilisation and erosion control would require characterising the effects of erosion (and other forms of soil degradation) on agricultural productivity. It is difficult, however, to draw a direct connection between a quantity of non-eroded (stabilised) soil and a quantity of agricultural product not produced. Adopting the replacement costs method requires defining the different substitution techniques that may be used to maintain yields, keeping in mind that the most suitable conservation practices will vary depending on the degree of erosion risk present in a given field.

Economic assessment of ES provided to society as a whole

This ES also directly benefits society as a whole by helping to prevent mudslides and land slippages and by improving surface water quality through the reduction of waterborne particulates. In theory, an economic assessment of this ES could be obtained using the replacement costs method, taking into account the cost of the associated water treatment processes for domestic, industrial, or recreational use. The avoided damages method could also be used, recognising that some of the economic value of this ES corresponds to the (avoided) cost of repairing buildings or other infrastructure impacted by mudslides. In practice, however, in addition to the difficulties cited above with respect to the economic assessment of other ES, large-scale implementation of these methods is challenging. The effects of erosion can vary significantly depending on both the ecological and the socioeconomic characteristics of the watersheds in question, and thus cannot be characterised in a general fashion. Revealed preferences methods are thus not that useful for assessing ES provided to society as a whole at the national level.

6. Towards the management of ecosystem services supply

Assessment of the current level of supply of different ES provides key information for the development of management strategies for ES supply in response to local or global issues. A single geographic unit (field, SAR, watershed, etc.) can provide multiple ES, some of which will be based on shared biophysical determinants and/or be influenced by common external factors. As a result, any change in ecosystem management (notably through changes in agricultural practices) has the potential to affect the supply of multiple ES. Considered from the perspective of a single ES or exclusively from the perspective of the production of agricultural goods, ecosystem management becomes a question of maximising that ES or that agricultural good, potentially to the detriment of one or several others. An emblematic example of this is the maximisation of agricultural goods from agroecosystems based on the use of external inputs, to the detriment of biological diversity, the foundation of all ES.

To develop a strategy for ES management at the territorial level, it is thus necessary to move from the analysis of individual ES to the integrated analysis of multiple ES. A multi-services approach seeks to identify the various ES currently supplied to the farmer and to society, the ways in which these ES interact, and the levers available to protect or enhance all or several of these ES together.

Strategies and policies for ES management must then be articulated with those focused on other objectives, such as the conservation of biodiversity or the minimisation of agriculture's environmental impacts.

From the analysis of individual services to a multi-services approach

Historically, most work on ES has focused on the assessment of one or a small number of ES. One meta-analysis found that 50% of such studies examined a single ES in isolation, without considering interactions with other ES. In recent years, a growing number of multi-service analyses have been undertaken. Work of this type examines relationships among ES either in terms of biophysical factors (ES supply) or in terms of the uses that are made of ES or the preferences of relevant actors (ES demand). In EFESE-AE, relationships among ES were considered exclusively in terms of ES supply, following to two complementary approaches: i) analysis of ES supply at the territorial level (ES bundles); and ii) the identification of functional relationships among ES (interactions). While the analysis of ES bundles offers a diagnostic tool for assessing average levels of ES supplied within a given territory, it is insufficient for developing actions intended to manage or modify ES supply.

▌ The identification of bundles of ecosystem services

A bundle of ES is defined here as a set of observed ES within a given spatial unit over a given period of time. The "shape" of the bundle is determined by the respective levels of supply of the ES belonging to it, although the underlying terms of their coexistence may not be known. Thus, two ES may be found together within a single spatial unit without directly interacting. A variety of methods exist to analyse the spatial co-occurrence of ES supply. Note that the analysis of ES bundles is usually static; research on the temporal dynamics of ES bundles is just beginning to appear in the literature (cf. Box 6-1).

When implemented at the level of a specific territory, the analysis of ES bundles can assist decision-makers in developing a "diagnosis" or assessment of ES supply, a precursor step toward the establishment of objectives for the shape of the ES bundle to be provided (ideally) by the territory. This approach currently still constitutes a methodological challenge.

Analytical methods for the spatial coexistence of levels of ES supply

The analysis of ES bundles is an extension of the approach known as "by pairs," which consists in analysing the spatial co-variation of ES levels two-by-two using correlation coefficients and indices of spatial overlap. Extended to the study of the levels of supply of n ES within a given spatial area, this approach relies in particular on the analysis and multivariate classification of spatial units characterised by the average levels of ES they supply. Principal Component Analysis (PCA) and Pearson correlation coefficient are the two most frequently used quantitative methods for identifying correlations between ES. Data-clustering methods allow for the identification of groups of spatial units sharing the same pattern of ES levels, or in other words, the same shape of ES bundles. The methods of cluster analysis most often used in the study of ES are hierarchical cluster analysis and "k-means" clustering. As an extension of the latter, the use of self-organising maps (based on artificial neural networks) provide an interesting alternative for examining and mapping the spatial distribution of complex, multi-dimensional datasets.

In general, an analysis of ES bundles consists of six steps:
1. Quantification of each ES, analysis of the consistency of the results, and identification of the scope of application and the limitations of the individual assessment (see Chapters 2, 3, and 4);
2. Choice of criteria for the construction of bundles (e.g., by beneficiary type) and for the level of spatial resolution for the bundle analysis;
3. Aggregation (or, more rarely, disaggregation) and standardisation of values for each ES at the spatial level of the bundle analysis and analysis of the results of the change in scale;

4. Assessment of paired relationships between ES at the same spatial level as for analysis of the ES bundles (optional, but facilitates analysis of the results in the following steps);

5. Classification of spatial units as a function of the shape of their ES bundles, and analysis of the shapes of bundles by class;

6. Analysis of the relationship between ES bundles and biophysical determinants or external factors.

Box 6-1. The temporal dynamics of ES bundles

Analysis of the temporal dynamics of ES bundles requires determining the dynamics (inter- or intra-annual) of biophysical determinants and external factors. Understanding these dynamics is particularly important given that ES assessment is often undertaken within the context of climate change and/or changes in land use. In existing assessments, levels are estimated either by using historical data or based on predictions in the context of a modeling approach. To date, two types of strategies have been used to analyse the dynamics of ES bundles under different scenarios: the matrix approach, used to follow the effects of land-use changes; and "state and transition" models, used to analyse and represent the effects on ES bundles of changes in ecosystem condition.

In the matrix approach, the value for ES supply is defined by an assessment based on expert opinion, or on spot observations, for each land use category, and then remains unchanged regardless of the scenario. As a result, the prediction of ES bundles under different scenarios is based exclusively on changes in the spatial configuration of different types of land use within the area under consideration.

For "state and transition" models, the goal is to represent the link between the dynamic of ecological "states" (e.g., transformation of a meadow into a wasteland) and the dynamics of the ES bundles associated with these different states. Approaches based on "state and transition" models can make use of expert opinion, observations and/or measurements in the field and modeling of the mechanisms involved in the supply of these ES.

Challenges associated with the analysis of ES bundles

Accounting for multiple ecosystem types

To be useful for decision-making purposes, the analysis of ES bundles should be made at a spatial scale that is relevant to territorial planning issues, public policy implementation, and/or ecosystem management. In most cases, multiple ecosystem types (forests, wetlands, urban areas, various types of agricultural ecosystems, etc.) will coexist within the area to be considered. For each ES, a set of indicators corresponding to the specific characteristics of each ecosystem type present is thus needed. Some ES are assessed using "generic" variables (e.g., the ES for global climate regulation, evaluated through the quantity of C stored in the soil, regardless of the type of ecosystem). For other ES, however, it may be necessary to identify specific indicators based on the nature of the ecosystem (e.g., the ES for pest regulation, assessment of which is based on landscape composition indicators defined with respect to agricultural ecosystems).

Much of the current work on ES bundles assessment elides this difficulty by assigning a "zero ES" level for ecosystem types that are incompatible with the chosen indicator. Although this simplifies the methodology, it fails to distinguish situations where the ES is truly zero from those where the ES level is unavailable due to the lack of a suitable indicator.

Another way to address this problem is to conduct an analysis of ES bundles by ecosystem type, or even sub-type (e.g., one analysis of ES bundles supplied by major field crop ecosystems, another of ES bundles supplied by grassland ecosystems). This option present several other methodological issues, however. For example, although on the one hand the ES bundles analysis should be undertaken at the level of the entire country, at a specified resolution (spatial unit), a focus on ecosystem type implies selecting only those spatial units most relevant to the analysis. To make this selection, specific criteria and thresholds must be defined to identify which spatial units to select. These criteria and thresholds must be defined all the more carefully in situations where the supply of an ES by an ecosystem depends on the characteristics of the surrounding landscape, and thus on other types of ecosystems, which are nevertheless excluded from the analysis. Methods for determining these thresholds relate to research questions that were beyond the scope of EFESE-AE.

Applying change-of-scale procedures to chosen indicators

To conduct an analysis of ES bundles at a given level of spatial resolution, all the ES to be examined must be quantified at the same scale. If they are not, change-of-scale procedures (aggregation or disaggregation) must be applied to the indicators so as to estimate an "average" level of supply for each ES at the spatial resolution chosen for the ES bundle analysis. Depending on the methods employed, however, the change of scale can have a significant effect on the estimated level of ES supply. The average level of ES supply at the scale of the SAR, for example, may correspond to highly contrasting values at the resolution of the field parcel. Methods for determining appropriate change-of-scale procedures and their effects on ES bundle analysis correspond to research questions. An evaluation of the effect of various change-of-scale procedures on the "average" level of ES supply, and on the identification of ES bundles, may be performed as a type of sensitivity analysis. This could be accompanied by a broader reflection on the comparative assessment of goods and services at different scales and the identification of corresponding ES bundles.

Demonstration of the potential of ES bundle analysis

The assessment of individual ES should be completely stabilised before undertaking an ES bundle analysis. As indicated in Chapters 2, 3, and 4, however, some ES indicators present methodological limitations and/or only imperfectly represent the level of

ES supply. In addition, the scope of the assessment (number of different types of agricultural ecosystems involved) varied depending on the ES.

Due to the organisation of the EFESE-AE project, the evaluation of certain ES was not stabilised at the time when the analysis of ES bouquets was launched. This analysis was carried out as an illustration, and the results it provides should not be used for their own sake. The ES bundle analysis was applied to major field crop ecosystems at the level of the SAR, using intermediate results from the biophysical assessment of the ES. Only those SAR with a significant area of agricultural land planted to major field crops were retained for the analysis.[53] Two ES bundle analyses were conducted in parallel: one for ES supplied to the farmer, the other for ES supplied to society as a whole. Each analysis thus had its own list of ES/goods, although the two lists were not mutually exclusive (since some ES have both types of beneficiaries). For each ES, the mean value was estimated at the level of the SAR, and then the mean ES values for each SAR were standardised. Next, a correlation analysis of ES levels vs. crop production output (goods) was completed. Finally, a typology (classification) of the shape of the two ES bundles (on re: the "farmer," the other re: "society") at the level of the SAR was determined using self-organising maps. This classified the SAR into homogeneous sub-groups (or clusters) sharing the same characteristics in terms of the co-occurrence of levels of ES/agricultural goods within the SAR.

Identification of correlations between ES levels

Several representations of the correlations among ES levels and the production of goods were tested. A network of correlation (Figure 6-1) provides an intuitive graphic representation of the direction and degree of correlation among the variables, making it possible to identify groups of variables more strongly linked among themselves. This type of analysis thus provides an initial representation of the co-variation of levels of ES supply, considered at the scale of all the spatial units (in this case, SAR) included in the analysis.

Keeping in mind that intermediate results were used for this analysis, this correlation network makes it possible to identify three groups of goods and ES that are more strongly correlated:

- One group includes the indicators for the input ES for supply of mineral N and return of water to crop plants, total level of plant production, and the ES for water quality regulation with regard to N, all of which are positively correlated;
- A second group includes the indicators for ES linked to the presence of semi-natural elements or to the nature of plant covers: ES for soil stabilisation, the recreational potential of agricultural ecosystems, crop pollination, climate regulation, and the regulation of pest insects, all of which are positively correlated, and the ES for the

---

53. An arbitrary threshold was chosen for illustrative purposes: SAR in which the land area in major field crops was more than 33% of the total land area in major field crops or grasslands.

regulation of weed seeds by carabids, which is negatively correlated with all the other ES in this group;

• A third group made up of the stock of C in the soil and the return of blue water, which are positively correlated, and annual storage of C in the soil, which is negatively correlated with these two ES.

This type of analysis enables us to see how one ES indicator is isolated within the correlation network or how another indicator is positioned in between two distinct groups of ES. In addition, this representation makes it possible to identify indicators that are negatively correlated with a group of other indicators, which are positively correlated among themselves. The analysis of the functional significance of these relationships was not taken any further, given the illustrative nature of the results.

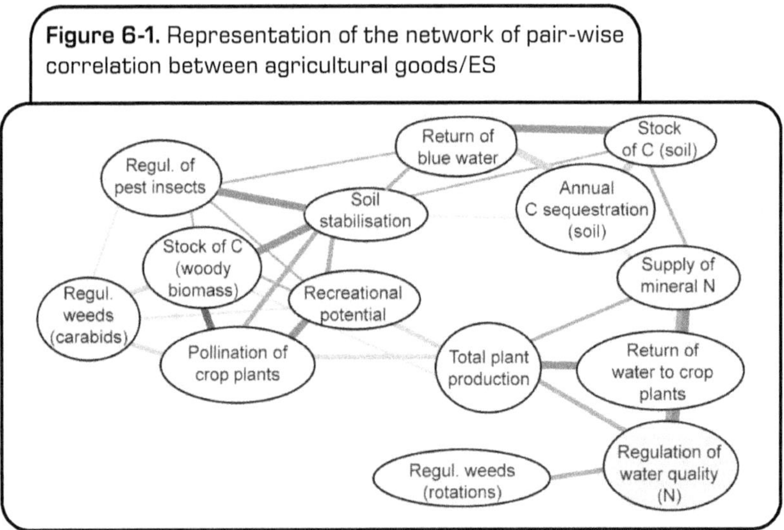

Figure 6-1. Representation of the network of pair-wise correlation between agricultural goods/ES

The analysis of ES bundles and agricultural goods

The analysis of ES bundles was conducted in two steps: i) identification of groups of SAR with similar profiles in terms of the supply of agricultural goods and ecosystem services (that is, showing the same shape of ES bundles) by type of beneficiary; and ii) analysis of the spatial congruence of the different bundle shapes associated with the two types of beneficiaries.

Figure 6-2 shows a possible summary illustration of the key findings from these two steps. The radar or spider charts show the shape of the bundles that have been identified for the two types of beneficiaries. The analysis of the "farmer" bundles reveals three groups of SAR, each of which is characterised by a distinct bundle shape. Similarly, four types of "society" bundle shapes, each associated with a group of SAR, were identified. In the radar charts the level of each of the ES is standardised and varies between 0 and 1. The representation of the distribution of bundle shapes across SAR makes it possible to observe the spatial congruence of ES/goods for each type of beneficiary.[54]

For example, for the dataset used here (for the purposes of illustration), the ES bundle "A" provided to farmers can be described as follows:
- A1: high levels of ES, with the exception of the ES for the regulation of weed seeds by carabids;
- A2: intermediate ES levels overall (on average, between A1 and A3), with the highest level of regulation of weed seeds by carabids and the lowest levels of regulation of pest insects, soil stabilisation, and crop pollination;
- A3: the lowest levels, on average, for the ES for storage and return of water to crop plants, regulation of weed seeds, and total plant production, but with the highest levels for soil stabilisation and the regulation of pest insects.

In the second step, the analysis of the spatial overlap between the four forms of the "society" bundle and the three forms of the "farmer" bundle is shown in a contingency table for the SAR (number of SAR at the intersection of each possible pair of "farmer" and "society" bundles). As in the preceding step, the congruence of the "farmer" and "society" bundles was also represented spatially. For the test dataset used here, a strong congruence was observed between clusters A2 and S2, clusters A3 and S1, and clusters A3 and S4 (see Figure 6-2).

This type of analysis – of the congruence of ES bundles supplied to different types of beneficiaries – can be used to characterise the multi-functionality of the spatial units (e.g., provision of multiple ES to farmers and to society) or, alternatively, their specialisation.

Methodological limitations

With self-organising maps, the assignment of a bundle form to an SAR is linked to the SAR's spatial position. A possible line of inquiry would be to identify SAR linked to different ES bundles shapes over the course of the iterative process as a way of gauging the uncertainty of the assignment of bundle shapes to SAR. It would also be interesting to explore the use of fuzzy clustering methodologies to make these assignments.

Generally speaking, a number of different methodologies and approaches can be used to identify the shape of ES bundles, each with its various advantages and disadvantages. The choice of spatial resolution, ecosystem type, ES and goods to be considered, analytical methods and the way which results are to be presented should all be determined with respect to the objectives of a given analysis. For example, if one is seeking to determine the relative impact of management and landscape composition on the shape of ES bundles, analyses at the level of the watershed, territorial development unit, or SAR are likely to be most relevant.

---

54. The shape of a "farmer" bundle cannot be compared to the shape of a "society" bundle.

Figure 6-2. Example of the types of results that can be obtained using an analysis of bundles of ES/goods

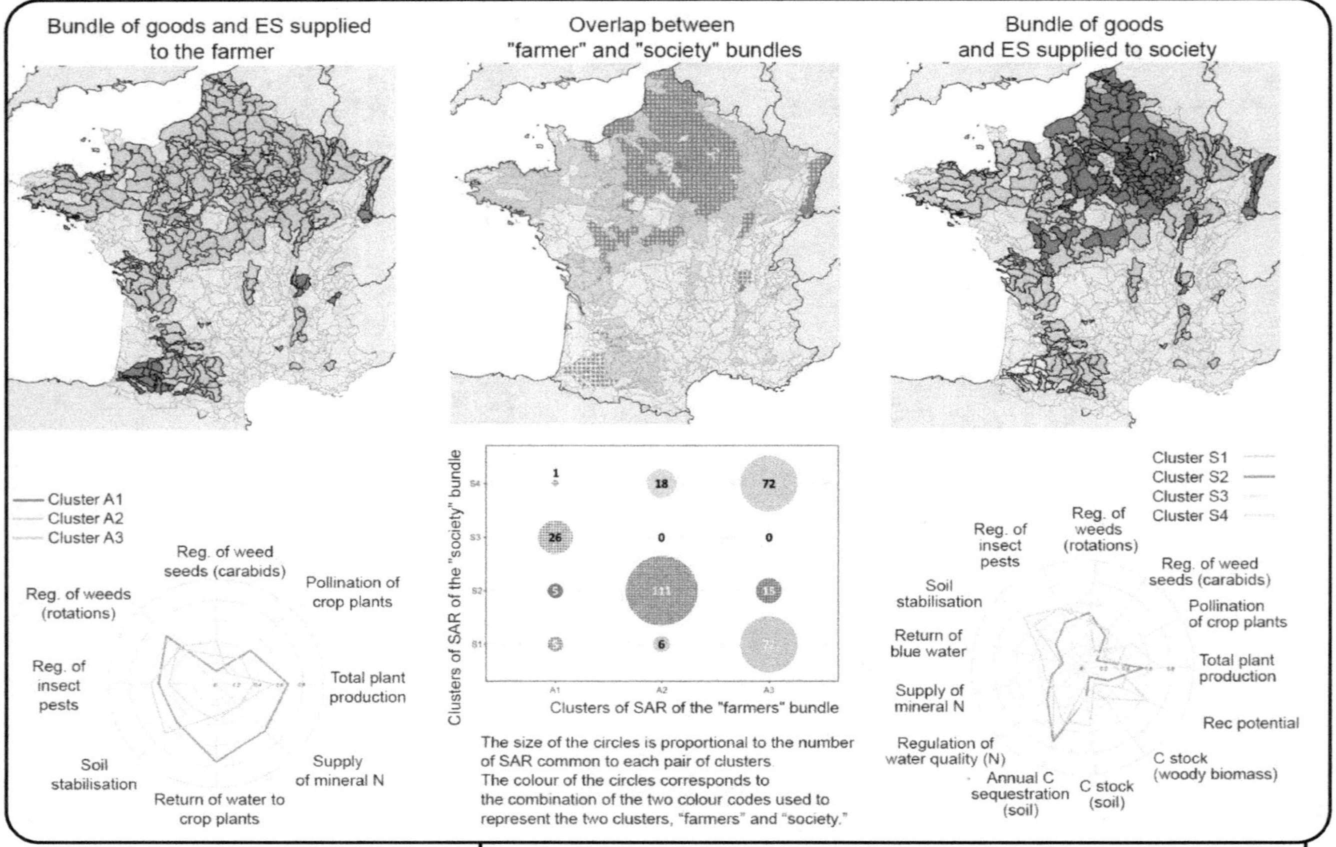

Feedback loops between ES and biodiversity and ES and organic matter are not shown.

▌ From biophysical determinants to interactions among ecosystem services

Methodologies for the analysis of ES bundles do not make it possible to infer the co-variation of ES over time, since they do not provide information on the ecological mechanisms underlying ES synergies or antagonisms.[55] Undertaking a rigorous analysis of ES interactions requires a detailed knowledge of the biophysical relationships existing among ES.

Given the diverse range of biophysical processes involved in the supply of the ES examined in EFESE-AE, each ES was described by the expert(s) whose disciplinary field was most relevant (ecology, agronomy, hydrology, animal science, etc.). As a result, the final list of biophysical determinants for the different ES could only be established once all the ES were fully described, and once the concepts and terms specific to each discipline had been harmonised. It was then possible to identify the "key" determinants influencing the supply of multiple regulating ES examined.

Because biodiversity is central to both policy discussions and management strategies relating to ES, the analysis focused on aspects of biodiversity as a key determinant of several ES. It is important to keep in mind that the biophysical determinants identified here are specific to the list of ES examined in EFESE-AE (which did not include all ES).

Figure 6-3 which results from this transversal analysis reflects the fact that agricultural ecosystems are complex systems, characterised by numerous interactions among different entities. In the interests of simplification, the figure does not show feedback loops running from ES back toward the biodiversity components shown here, nor towards other biodiversity components that are not shown (e.g., wild flora or fauna in the landscape). Although it is a simplified representation with regard to numerous existing interactions, the figure was created to help identify the principal "targets" for strategies of agricultural ecosystem management seeking to strengthen ES provided to ecosystem managers and to society.

Given their number and their underlying complexity, it is difficult to describe all the relationships presented in the figure. However, the figure highlights the importance of:
- the spatiotemporal distribution and the diversity of managed plant covers (crop covers, weeds present in the field and semi-natural elements within field boundaries, such as hedges and isolated trees);
- the abundance and diversity of three components of associated biodiversity: crop auxiliary species (pollinators, pest predators); endogenous meso- and macrofauna; and soil microorganisms;

---

55. In the international literature, the terms *trade-off* and *synergy* are used to describe: i) ecological (biophysical) interactions among ES; ii) temporal variations in ES supply; iii) the spatial co-occurrence of ES; iv) biophysical interactions resulting from ES management; v) disjunctures of supply and demand; vi) compromises between costs and benefits; and vii) compromises among different beneficiaries. In EFESE-AE, *antagonism* is used to designate an antagonistic biophysical interaction between ES, and *compromise* to designate situations in which social negotiations are involved relative to a demand for ES or methods of ES management.

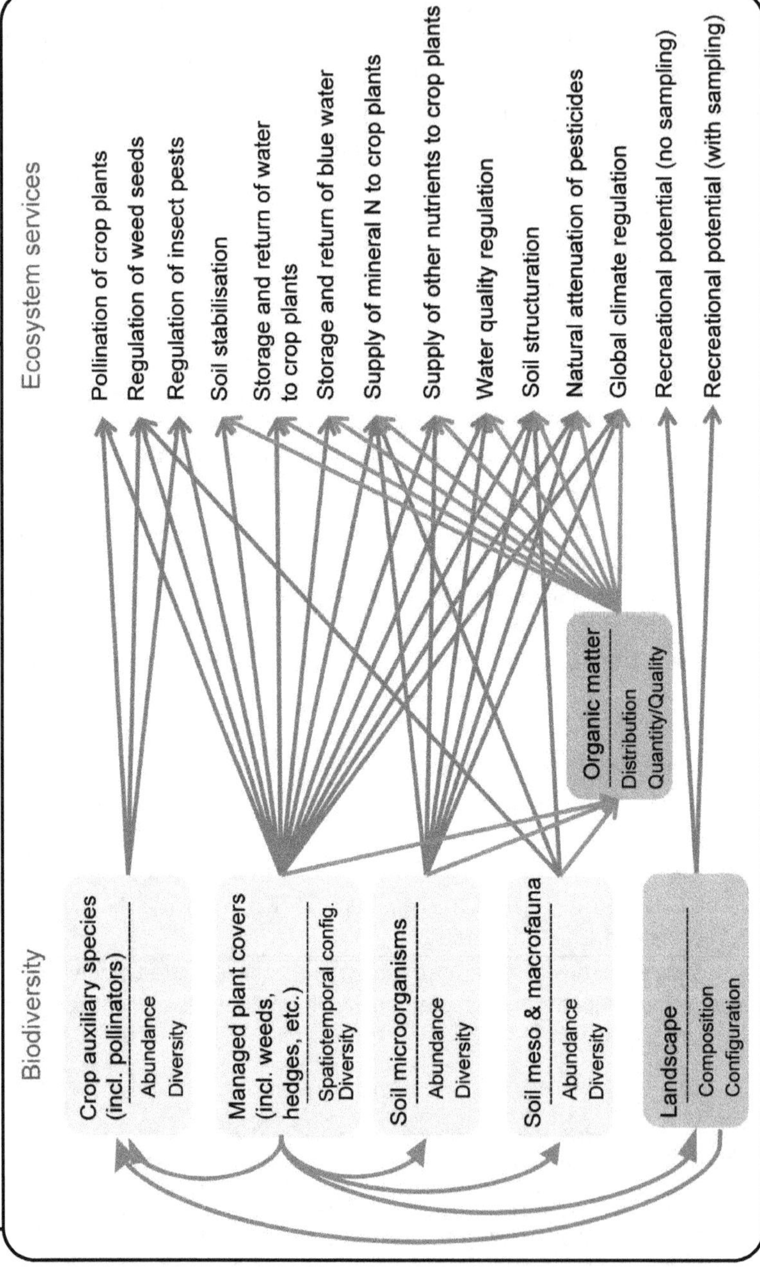

Figure 6-3. Principal interactions among ES through biodiversity components

- the quantity, quality and distribution of organic matter in the soil;
- landscape composition and configuration.

The six main types of biophysical determinants were established for the 14 ES retained in the study (12 regulatory ES and two cultural ES).

It is important to note, first, the central position and role of the spatiotemporal configuration of managed plant covers within the field parcel. In addition to crops, these can include covers established for purposes other than agricultural production (e.g., strips of grass or flowering plants), as well as associated managed plant biodiversity such as volunteers, weeds and semi-natural areas present within the field area. Managed plant covers have a direct impact on the provision of 11 of the 12 regulating ES examined here, and an indirect impact on the 12th (natural attenuation of pesticides) through their effects on the structure and abundance of microbial communities. This indirect effect involving soil microbial communities also determines the level of supply of five other regulating ES, such as the ES of supply of nutrients (N, P, etc.) to crop plants, soil structuration and global climate regulation. Managed plant covers also strongly determine the composition and configuration of agricultural landscapes, and thus ES that directly depend on these landscape features (recreational potential). Moreover, in addition to their direct effects (so-called bottom-up regulation), they also indirectly impact the three ES of biological regulation through their effects on the abundance of crop auxiliary species communities (top-down regulation). Finally, managed plant covers determine the structure and abundance of soil meso- and macrofauna, and thus have an impact on ES that depend on species communities, both directly (soil structuration) and indirectly through organic matter (8 regulating ES).

If managed plant cover plays a major role in the constitution of the organic state of the soil, the microbial communities as well as the meso- and macrofauna of the soil are not left out. Here again, methodological advances were proposed through the use of dynamic modeling of soil-plant(-animal) systems. Dynamic modeling allows one to account for interactions between managed plant covers (crop rotations), the organic condition of soils and ES relating to the water, N, and C cycles. Few large-scale ES assessments to date have been based on these types of dynamic simulations.

The diagram in Figure 6-3 also assists in highlighting the different levels of organisation involved in the supply of the ES under consideration, particularly at the field level and the landscape level. ES relying on soil biodiversity are provided by the soil-plant system present in the field. Certain processes linked to the lateral movement of water depend on watershed functioning (lateral hypodermic drainage, surface run-off) but it is the outcome of these processes at the level of the field parcel (the quantity of water involved) that contributes to the supply of input ES.

ES provided by aboveground and airborne fauna depend both on conditions within the field and the composition and configuration of the surrounding landscape matrix (notably the semi-natural elements), which provide habitat and food resources to these taxa: although these ES are expressed at the level of the agricultural field (or group of

131

fields), they are linked to the characteristics of the wider agricultural landscape (from one to several kilometers beyond the field limits).

We should note that the ES for soil stabilisation and erosion control, through the maintenance of soil characteristics, is strongly linked with ES for the supply of nutrients to crop plants, the storage and return of water to crop plants, and the regulation of water quality. The ES for soil structuration, through its effect on soil structure, also has a key role. The ES for the supply of nutrients to crop plants interact more particularly with the ES for the regulation of water quality with respect to N, P, and DOC.

This schematisation of ecosystem functioning and interactions with the landscape, provides an initial perspective on the key ecosystem components farmers (through agricultural practices) and other landscape managers should focus on in order to protect or modify levels of ES supply. It underscores the major role played by the spatiotemporal configuration of agricultural ecosystems (and landscapes), and thus the centrality of practices that affect these spatiotemporal patterns. It helps us grasp the potential effects on ES of changes made to ecosystem components, whether intentional or unintentional. It shows clearly, and in an integrative fashion, the large number of indirect interactions among ES, and thus the need to develop decision-making tools for the sustainable management of ecosystem conditions and associated ES. Such tools would make possible a more precise knowledge of the relationships between ES according to local pedoclimatic conditions, and would perhaps make it possible to anticipate the possible effects of climate change.

This analysis thus enables us to go beyond general approaches based on "generic" indicators for the maintenance of the good ecological condition of ecosystems. It can provide a framework for the design and implementation of field-based monitoring programs focused on biodiversity and other environmental factors (such as the French National Biodiversity Observatory), examples of which are currently in development.

▌Identifying potential levers for ecosystem services management

It is also important to identify operational levers allowing action on these variables taking into account the configuration of the ecosystem and the landscape. These levers correspond to anthropic factors external to the ecosystem, which, through their impacts on biophysical determinants, can influence ES levels. In the case of agricultural ecosystems, these external anthropic factors include agricultural practices for the management of soils and plant biomass (four major categories of practices). These practices were not ranked in terms of the magnitude of their effects on key biophysical determinants. Determining the precise impact of each of these practices on ecosystem characteristics would require detailed work to categorise their nature and intensity (e.g., modes and intensity of soil tillage) followed by an analysis of how their interactions with ES levels. However, following the same approach as for the identification of key ecosystem and landscape components, a transversal analysis of

Figure 6-4. Principal indirect interactions, at the field level, between external anthropic factors and regulating ES

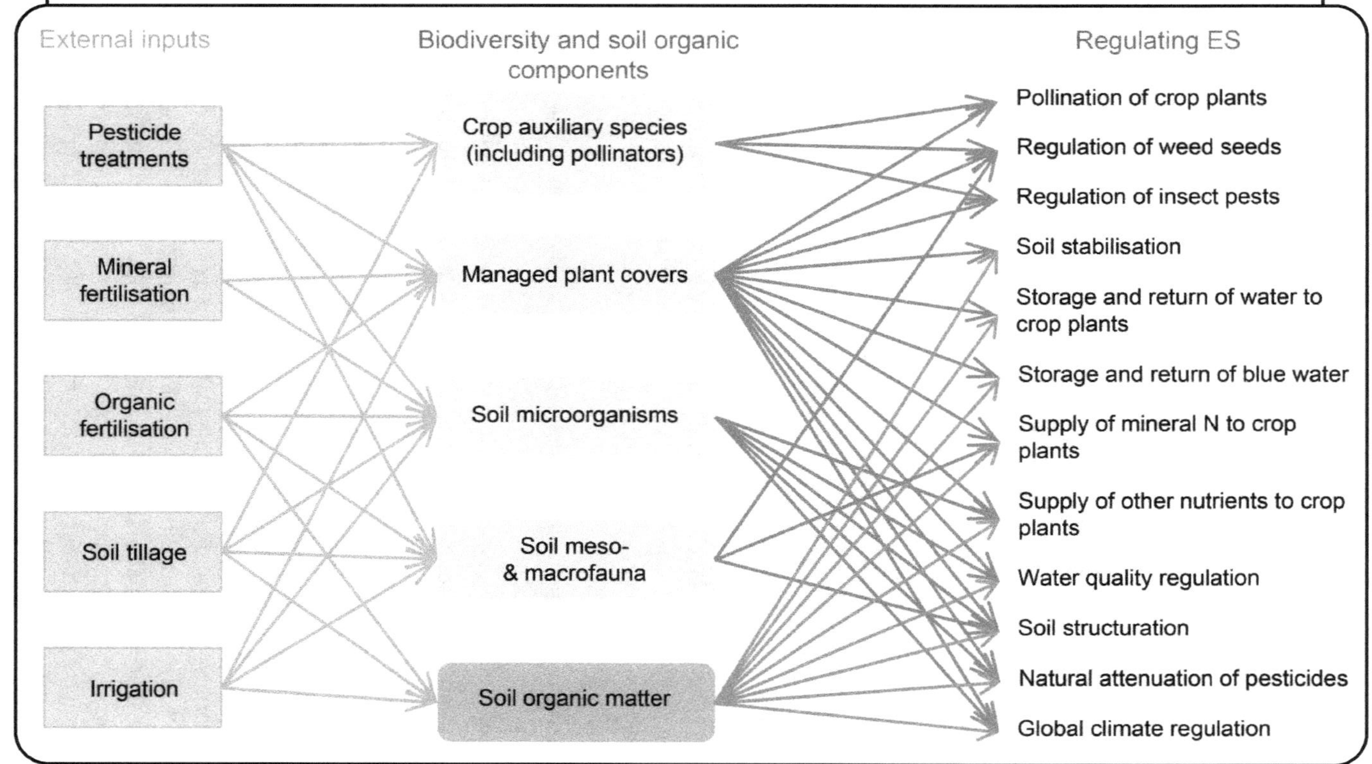

external factors impacting each ES allowed for the identification of the major types of practices that could act as management levers, and their status with respect to ES.

Figure 6-4 offers a representation of the indirect connections between external agricultural practices and ES through their biophysical determinants. For the sake of simplification, the figure does not show feedback loops between ES levels and external anthropic factors through the activities and behavior of the farmer. Finally, we should note that although climate may also be considered as an external factor acting on all agricultural ecosystem components (see Chapter 1), it is not included in this figure.

Pesticide treatments affect ES of biological regulation through their impacts – usually negative – on the structure and abundance of crop auxiliary species, and on the plant hosts of these species, including some weeds. Pesticides also influence the expression of numerous ES through their effects on soil microbial communities and soil meso- and macrofauna. However, pesticides can have positive effects on the levels of some ES in some situations. For example, regular use of the same pesticide can boost the ES for natural attenuation of pesticides through selection effects on the microbial communities capable of breaking down the product. It should be noted that this type of crop protection strategy can also lead to the development of pest resistance to the molecules being applied.

Soil tillage has effects that are similar to those of crop protection treatments. It acts as a disturbance of the biological functioning of soil microbial communities and meso- and macrofauna, and thus can influence the levels of ES that depend on these populations. It can also act as a disturbance of crop auxiliary species move through the soil or which use the soil as a nesting site. Finally, it plays a key role in the distribution and dynamics of soil organic matter.

Mineral and organic fertilisation practices have an effect on the growth of plant covers. Organic fertilisation especially will influence the dynamics of soil microbial communities, meso- and macrofauna and the characteristics of soil organic matter.

Irrigation, through changes in soil moisture levels, influences the growth of plant cover, the functioning of soil biological communities and the dynamics of soil organic matter.

Ecosystem services, biodiversity conservation and environmental impacts

Management levers previously identified will not necessarily be the same for other types of ES and/or ecosystems, or given different objectives on the part of decision-makers. The following analysis seeks to examine the connections between strategies and policies for ES management, on the one hand, and strategies and policies relating to biodiversity conservation or to the minimisation of the environmental impacts of agricultural practices, on the other.

▌ Ecosystem services supply and the conservation of biodiversity

Since the end of the 1990s and especially the beginning of the 2000s, the concept of biodiversity, at the heart of numerous scientific studies in ecology, has been increasingly associated with that of SE. Although the two terms are strongly connected, research on biological diversity and research on ES do not always have the same objectives. There is, in particular, a difference in underlying values: whereas biodiversity is often recognised as having intrinsic value, ES represent a recognition of utilitarian value. Scientific debate also exists as to the status of biodiversity within the conceptual framework of ES: e.g., should the maintenance of biodiversity through ecosystem functioning itself be recognised as an ES?

Mechanisms for understanding the relationship between biodiversity and ecosystem functioning: implications for ES management

The earliest studies on the relationship between biodiversity and ecosystem functioning date from the 1990s. Focusing on the effects of species diversity in grassland ecosystems, these studies found a positive correlation between grassland productivity and species richness. These initial findings have since been confirmed and generalised for other ecological processes, other levels of biodiversity, and other ecosystems. Review articles suggest that biodiversity, regardless of its level of organisation, is frequently associated with more effective ecosystem functioning. Such studies have also highlighted certain exceptions to this pattern, however, and in general have highlighted the need for further study of the interactions between biodiversity, ecological processes and ES.

Two major types of mechanisms underlying these interactions between diversity and ecosystem functioning have been identified (Figure 6-5). The identification of these mechanisms has direct implications for biodiversity management with a view to protecting ES:

- niche partitioning (or niche complementarity) among species, both spatial and temporal: different species or genotypes are likely to make use of different resources or different habitats, allowing for a better overall use of available resources. This effect increases as the number of different ecological niches increases, regardless of whether these niches are occupied by different species or genotypes;
- the sampling effect (also known as the selection effect): the greater species diversity, the more likely it is that highly productive species or individuals will be present. Moreover, if these highly productive species or individuals determine overall ecosystem productivity (they are competitive, and not excluded by less productive individuals or species), then one will automatically expect a positive correlation between diversity and productivity.

From the purposes of ES management, it is important to be able to distinguish between niche partitioning and sampling effects. The former imply that all or nearly all existing biodiversity is critical to the proper functioning of the ecosystem, while the latter imply that, at least under certain conditions, one or a few species (or genotypes) are most critical for proper ecosystem functioning. One result of niche partitioning is that certain mixtures of species can be more productive than the single most productive species alone (an effect referred to as transgressive overyielding). This phenomenon is not seen with the sampling effect, where the mean productivity of the species mixture will lie somewhere between the mean productivity and the maximum productivity of the individual species. Statistical methods can be used to separate partitioning effects from sampling effects based on data for the productivity of individual species/genotypes grown together vs. the same species grown in pure stands. A recent meta-analysis of studies of this type found that in general, both effects are present, with partitioning effects tending to be more important in terrestrial ecosystems.

Figure 6-5. Schematic representation of the partitioning effect (left) and sampling effect (right)

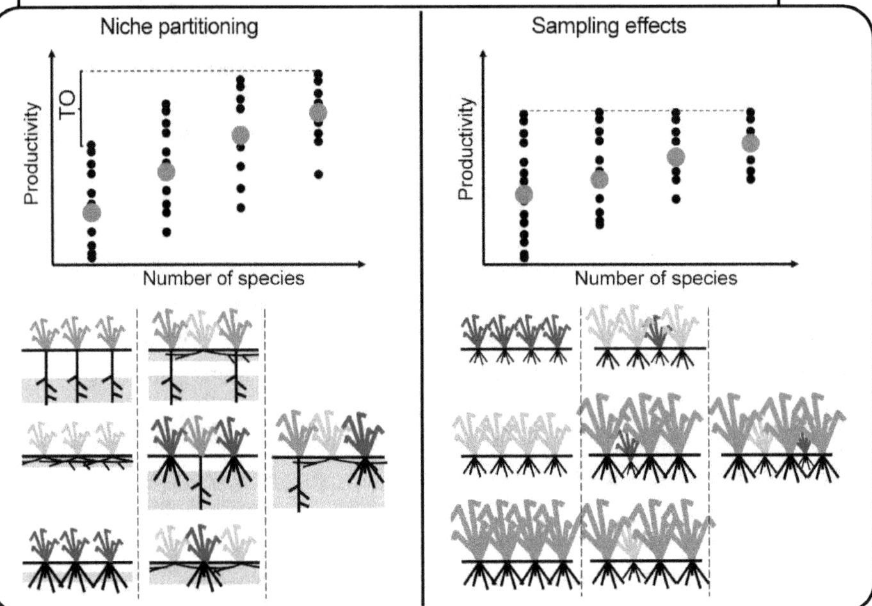

The red circles indicate mean productivity values as a function of the number of species. Niche partitioning among species, represented here by differences in rooting depth/rooting pattern among plant species (soil horizons explored by roots are shown in gray), allows for reduced competition and a better overall use of available resources. With sampling effects, the probability that a community will include highly productive species increases with the number of species present. If these highly productive species determine total productivity, there will be a positive correlation between species richness and total mean productivity.

Does the protection of ES implies the protection of (all) biodiversity?

The ES concept is presented in some studies as the only effective tool for ensuring biodiversity conservation. The increasingly important role granted to ES within conservation agendas, however, has given rise to considerable debate. One point of controversy relates to the risks of focusing too much on ES, while overlooking the true complexity of the ecological, social, and political challenges involved. The implications of advancing what is essentially a utilitarian view of nature have also been discussed at some length. Finally, some authors question the relevance of strategic choices for conservation: between the conservation of biodiversity and the conservation of ES, which is the end and which is the means? On the biophysical level, if we begin from the position that it is important to conserve all biodiversity, then the dilemma that presents itself is, what types of biodiversity are protected by protecting ES? This question may be addressed by considering two further questions: i) what types of biodiversity are essential to assuring the supply of all ES? (or, in other words, are some species redundant?) and ii) are there are compromises to be made between the conservation of biodiversity and the conservation of ES?

The share of biodiversity protected through the protection of ES depends above all on the number of species necessary to ensure the supply of a given ES, and thus, notably, on the relative importance of sampling effects vs. partitioning effects. If sampling effects are predominant, then ecosystem functioning and the supply of ES may be ensured by a very small number of highly effective species, and the conservation of biodiversity in general will require other arguments than those linked to the protection or the improvement of ES. The current consensus is that partitioning effects play the predominant role. Nevertheless, some level of functional redundancy among species/genotypes may be present, and can be evaluated based on the shape of the correlation curve between an indicator for biodiversity (e.g., number of species) and an indicator for ecosystem functioning (e.g., productivity, level of ES supply). In general, the slope of this curve will be relatively steep when diversity is low, and then tends to level off, suggesting the redundancy of some species. However, additional studies of grassland ecosystems have found that the species most important to ecosystem functioning can vary depending on the location, the moment, or the ecological process or pressure under consideration. Thus, even if certain species may appear redundant at a given moment or in a given location, all or nearly all grassland species are necessary to maintaining the full range of ES in environments subject to regular or irregular disturbances and variations. These results have been confirmed by more recent studies on a range of different ecosystems and taxonomic groups, suggesting that biological diversity is critical to the maintenance of ecosystem processes. In sum, these studies suggest that the majority of species are needed for the maintenance of ES at a given moment and over time, and that the percentage increases if one considers multiple ES and environments subject to regular or irregular disturbances and variations, with biodiversity helping to support ecosystem resilience (see below).

A few studies have begun to suggest that for certain ES considered individually, a limited number of species is sufficient to maintain the ES. In some cases, moreover, ES may be effectively supported by non-native species or domestic species. This can be true, for example, for biological control organisms that have become naturalised in agricultural ecosystems subsequent to their introduction; bio-control organisms may also help control other pest species, in addition to the targeted pest. In such cases, maintenance of the ES will not depend on indigenous wild biodiversity, which thus may be at risk of being neglected by a conservation program exclusively focused on ES.

Conservation efforts seeking to protect biodiversity, ES, and the production of agricultural goods may also have to contend with biophysical antagonisms between these different objectives. A few studies have found a positive correlation between ES supply and biodiversity, both at the local and the global level. On the other hand, many studies show a strong antagonism between agricultural production and biodiversity conservation. Recognition of this antagonism has given rise to debates about land use: should intensive agricultural production be concentrated within specific areas so as to reserve other areas for biodiversity conservation (a so-called land sparing strategy)? Or is it better to dedicate larger areas to less intensive forms of agricultural production that are compatible with greater protection of biodiversity (land sharing strategy)? The two strategies offer different advantages and disadvantages (between optimisation vs. better resilience). Choosing one strategy or the other will depend on the shape of the correlation curve between yield and a given biodiversity variable (e.g., species abundance). The nature of the correlation can vary significantly depending on the type of agriculture involved – agricultural systems based on external inputs vs. agricultural systems based on ES. Analysing these antagonisms and finding ways to move beyond the land sharing/land sparing debate will require assessing the effects of different types of agriculture on these antagonisms and the role of multi-scale effects (from the field to the landscape), of the spatial distribution of agricultural production and biodiversity conservation.

▌ The provision of ecosystem services and the management of agroecosystem environmental impacts

As described in Chapter 1, the study of ES provided by ecosystems and the study of the environmental impacts of agroecosystems offer two different types of information with regard to the functioning of the soil-plant system: information on one does not make it possible to directly infer information with respect to the other. Research completed in EFESE-AE provides an illustration of this fact. Thus, in addition to assessing the level of ES for water quality regulation and global climate regulation, the level of two key environmental impacts associated with these ES were also estimated: i) the amount of N leached beyond the ecosystem; and ii) net GHG emissions between the agricultural ecosystem and the atmosphere.

The link between cropping system impacts on drained water quality and its regulation by the ecosystem

The impact of "current" cropping systems on drained water quality was evaluated using two indicators estimated with the STICS model: the quantity of N leached, and nitrate (NO_3^-) concentrations in drainage water. On average across all the simulations, the quantity of leached N was 36 kg N/ha/yr, and the NO_3^- concentration was 54 mg NO_3^-/l/yr. The average annual values shown by these two indicators generally follow the same pattern of spatial distribution, with the notable exception of two geographic areas: the Landes and the foothills of the Pyrenees, on the one hand, and the northern part of the Saône Valley, on the other hand, show relatively high quantities of leached N but are not associated with higher NO_3^- concentrations than other geographic areas. A dilution effect of the annual quantity of drained water beyond the root zone may explain these particular findings (Figure 6-6).

Overall, these two environmental impact indicators are weakly correlated with the indicator for the absolute level of ES supply (quantity of non-leached N), and more strongly correlated with the indicator for the relative level of ES (proportion of non-leached N – cf. Figure 6-7). In all areas this correlation is negative, suggesting that the cropping system's impact on drained water quality tends to be lower where the ES level is high. The fact that this correlation is relatively loose suggests that the same capacity for N "retention" by the "soil-plant" system may be associated with very different impact levels. For example, situations in which 80% of the N entering the system is not leached are associated with quantities of leached N varying from 20 to 100 kg N/ha/yr, and NO_3^- concentrations varying from less than 50 mg NO_3^-/l to more than 150 mg NO_3^-/l. This result thus nicely illustrates the non-equivalent nature of the information provided by the two types of variables, ES and impacts.

Cropping system impacts on climate: GHG emissions

Cropping system impacts on climate were assessed using the annual net balance of GHG flows between the agricultural ecosystem and the atmosphere. The net annual flows of CO_2 and N_2O, weighted according to their respective GWP,[56] were estimated using the STICS model (flows of CH_4 were considered to be negligible for major field crops). Flows of CO_2 associated with the short C cycle (photosynthesis, autotrophic respiration) were not considered, since these are in large part self-compensating. The average annual net GHG balance therefore takes into account N_2O emissions and the storage/removal of C from the soil.[57]

Overall, results for the GHG budget from the simulations were in line with results published elsewhere for temperate cropping systems (Figure 6-8). A large majority

---

56. Global warming potential, 298 times higher for N_2O than for CO_2.
57. Without taking into account the storage/removal of C in woody formations associated with the agricultural ecosystem.

of cultivated agroecosystems are net emitters of GHG (positive values). On average, net emissions of GHG are approximately 1,029 kg CO_2e/ha/yr. This result is largely attributable to N_2O emissions, which are on the order of 1.9 kg N-N_2O/ha/yr on average. Areas with the highest emissions are found in the Southwest, in Poitou-Charentes, in Brittany, in Limagne and in Alsace. Only a small number of simulated cases, located in Beauce and in the northern part of the Camargue, are GHG sinks, a result of the low levels of N_2O emissions that characterise these systems.

Figure 6-6. Impact of major field crop system on water quality, estimated for cropping systems managed using prevailing agricultural practices

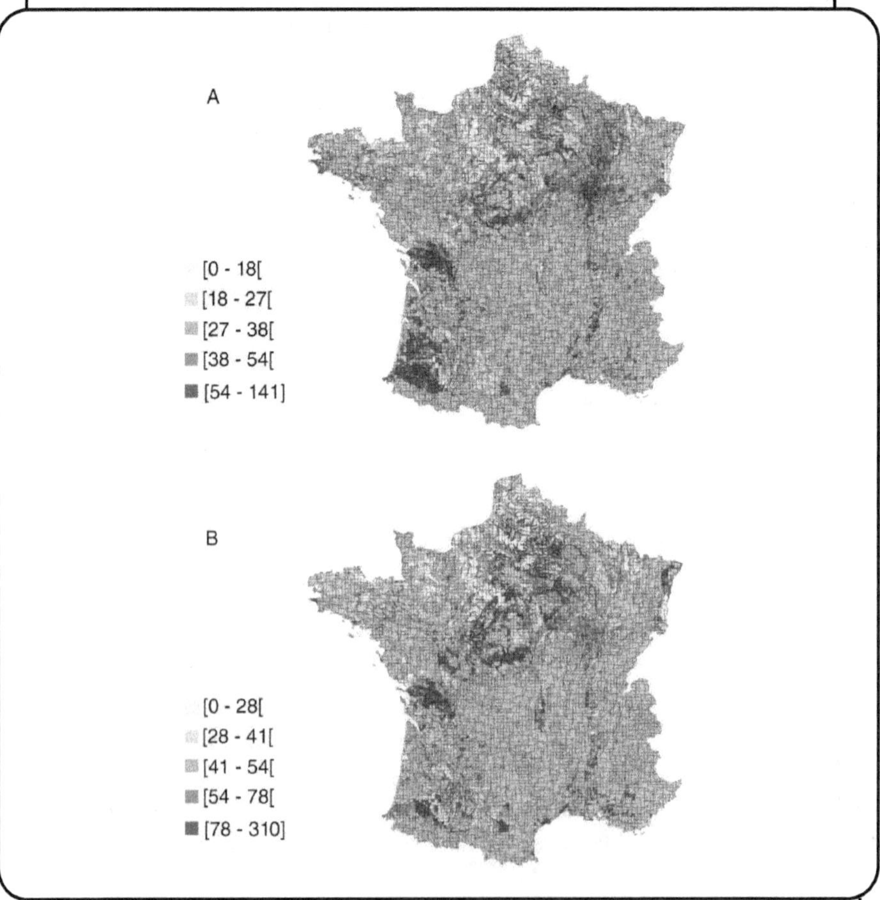

A: Average annual quantity of leached N (kg N/ha/yr); B: Average annual concentration of NO_3^- in drainage water (mg NO_3^-/l/yr); Spatial resolution: PCU; PCU in gray (including Corsica): no "major field crop" simulations; Value classes correspond to quintiles.

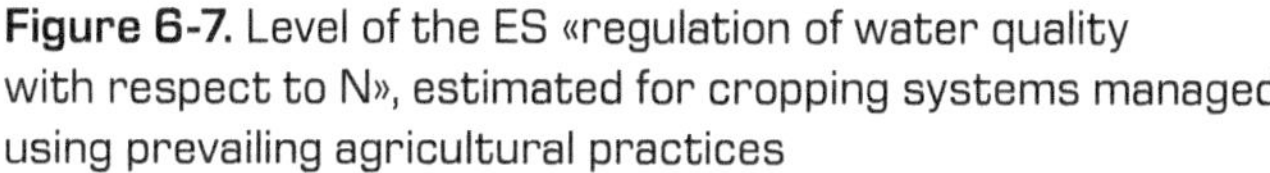

Figure 6-7. Level of the ES «regulation of water quality with respect to N», estimated for cropping systems managed using prevailing agricultural practices

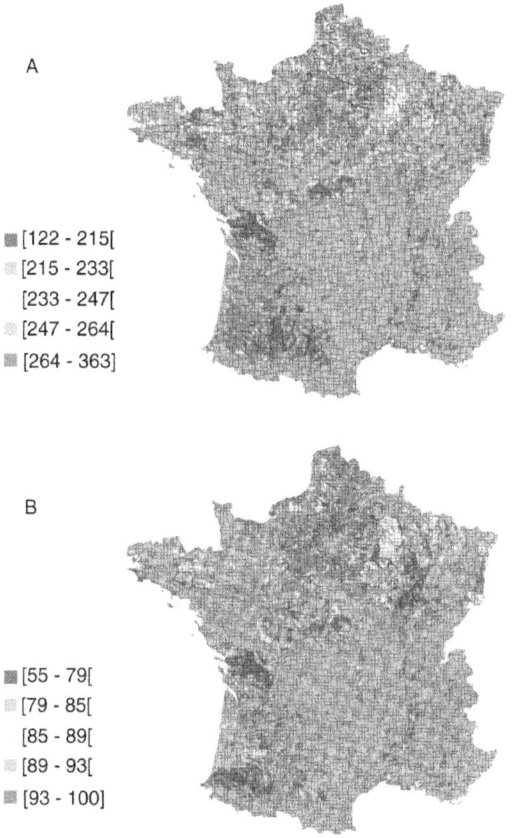

A

[122 - 215[
[215 - 233[
[233 - 247[
[247 - 264[
[264 - 363]

B

[55 - 79[
[79 - 85[
[85 - 89[
[89 - 93[
[93 - 100]

A: Average annual quantity of non-leached N (in kg N/ha/yr); B: Average annual percentage of non-leached N (in %). Spatial resolution: PCU; PCU in gray (including Corsica): no "major field crop" simulations; Value classes correspond to quintiles.

N_2O emissions tend to be higher where external inputs of N are higher, confirming the role of nitrogenous fertilisation in these emissions. The link is nevertheless a loose one because of the multiplicity of factors involved in the production of N_2O (temperature, moisture levels, pH, etc.). Finally, we can note the favorable effect of cover crops on GHG budgets (-130 kg CO_2e/ha/yr on average), by slightly increasing C storage in the soil (see Chapter 4) and slightly reducing emissions of N_2O.

Figure 6-8. Average annual net GHG emissions, estimated for cropping systems managed using prevailing agricultural practices

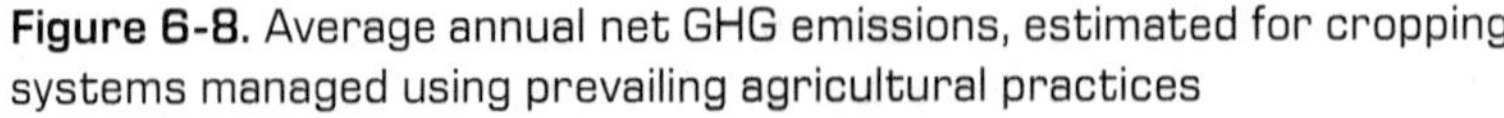

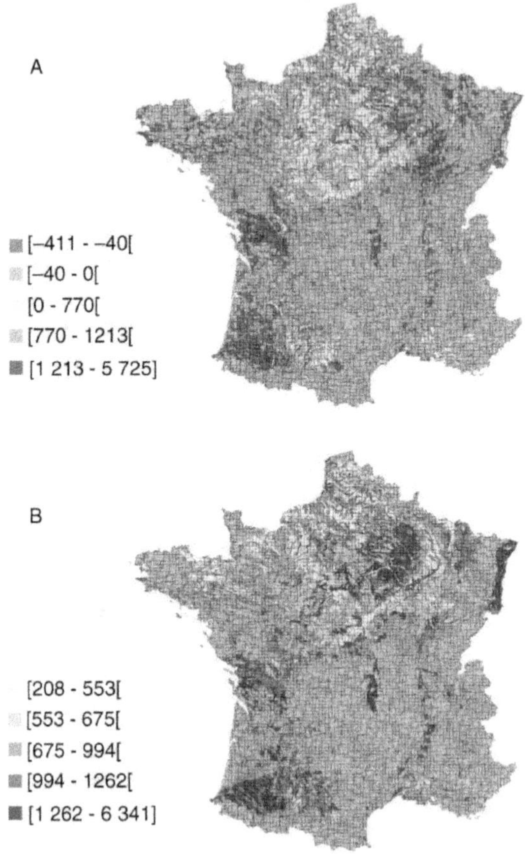

A: Average annual net GHG emissions (kg CO_2e /ha/yr), the range of values has been divided into five classes of the same size; B: N_2O emissions (kg CO_2e /ha/yr), value classes correspond to quintiles; Spatial resolution: PCU; PCU in gray (including Corsica): no "major field crop" simulations.

A qualitative review of the spatial distribution of the results shows that, while the level of the ES for global climate regulation (Figure 6-9) and net GHG emissions seem to be negatively correlated overall, a single level of net GHG emissions may be associated with different levels of ES and vice versa. Thus, for example, in the Landes, in the

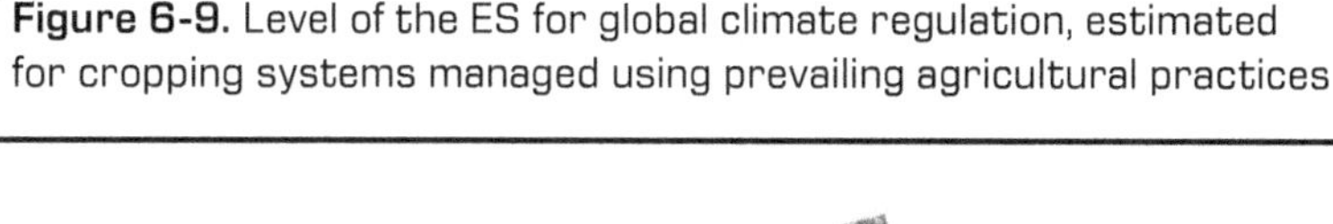

Figure 6-9. Level of the ES for global climate regulation, estimated for cropping systems managed using prevailing agricultural practices

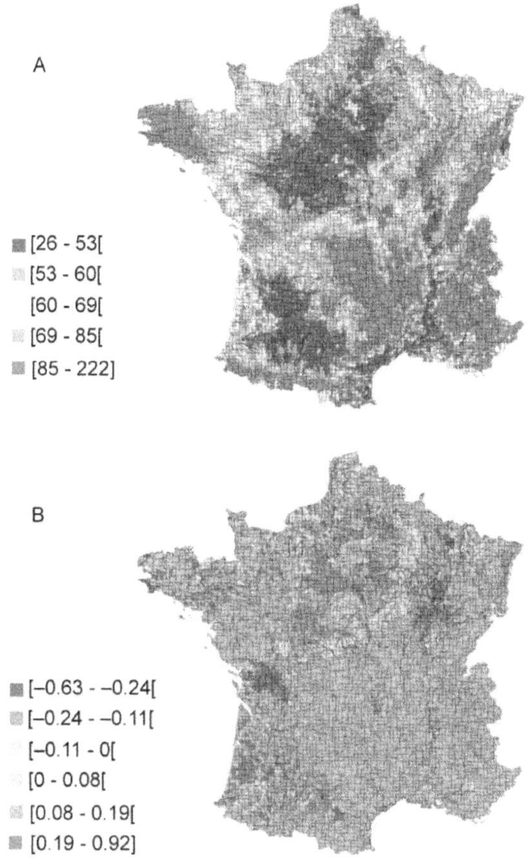

A: Total C storage in agricultural ecosystems (soil to a depth of 30 cm + woody material) in t C /ha, value classes correspond to quintiles; B: Mean annual C storage in the 0-30 cm soil horizon, estimated for "current" cropping systems, the range of values has been divided into five classes of the same size; Spatial resolution: PCU; PCU in gray (including Corsica): not estimated.

foothills of the Pyrenees, and even in Champagne, high net GHG emissions appear to be associated with both high and low levels of ES. In Limagne and in the southern part of Alsace, by contrast, the two indicators seem to be positively correlated, with high ES levels associated with high net GHG emissions.

7. Conclusions and perspectives

A framework for assessing the services provided by anthropised ecosystems

Agricultural ecosystems are a heavily human-impacted ecosystems, modified by farmers to produce agricultural goods. Among the key issues associated with understanding ES from agricultural ecosystems is the development of agricultural production systems based on ES, consuming few external inputs, and thus responding to societal issues. The analytical framework for assessing ES from agricultural ecosystems and the production of agricultural goods thus needs to help clarify the status and role of ES (linked to landscape functioning and the functioning of the "soil-plant-animal" system) vs. agricultural practices (including and external inputs) in supporting agricultural production. Consequently, consistent with the CICES, ES are defined here as ecological processes or elements of the structure of the ecosystem from which human derives benefits, sometimes by mobilising work, material or cognitive capital, in the aim of improving their well-being.

Accordingly, the analytical framework developed in EFESE-AE distinguishes between the biophysical determinants of ES, which are internal to the ecosystem, and other factors external to the ecosystem that can affect the level of ES supply as well as the level of agricultural production. This conceptualisation offers the possibility of assessing the ES potential of a given spatiotemporal ecosystem configuration and considering how that potential may be increased or decreased by external agricultural practices. A well-considered combination of the spatiotemporal organisation of plant cover and its management practices should meet objectives for agricultural production, the provision of ES to farmers (allowing a reduction in the use of external inputs), and the provision of ES to society.

Input ecosystem services provided to the farmer

The main purpose of the agricultural ecosystem is to produce plant (primary production) and animal (secondary production, developed from plant production) goods. Over the course of the cropping cycle, a certain number of ES have an impact on yield production by affecting the level of expression of factors that can limit or reduce yield, including water scarcity, nutrient deficiencies, insufficient pollination, and pest damage. Those regulating ES may thus be considered as factors of production, rather like external inputs (irrigation water, synthetic fertilisers, crop protection products, etc.). As managers of agricultural ecosystems, farmers are direct beneficiaries of these ES, referred to here as input ES.

Input ES can be divided into two major types based on their role in contributing to crop yields.

- ES that limit abiotic stresses help provide vegetative covers (grasslands or crops) with the conditions suitable for root development, including limiting water deficits and nutrient deficiencies: these are described as ES for "soil structuration," "supply of nutrients to crop plants," "storage and return of water to crop plants," and "soil stabilisation and erosion control." All of these ES are strongly dependent on the biotic and abiotic components of "soil" – soil fauna, aboveground and belowground plant systems, organic matter, available water reserve, etc. A central role is played by the ES for "soil structuration," which is itself a biophysical determinant of the other ES in this group.

- ES that reduce biotic stresses protect yields by limiting losses, such as those resulting from insufficient pollination or the activities of pest species: these correspond to "pollination of crop plants," "regulation of weed seeds," and "regulation of insect pests" ES. The level of these ES is strongly determined by the agricultural ecosystem's associated biodiversity. Supply of these ES thus depends both on the agricultural ecosystem in the strict sense (the field or parcel level) but also on broader landscape characteristics that help determine biodiversity dynamics.

▋ Key determinants and external factors for input ecosystem services

Figure 7-1 provides a schematic representation of the principal agricultural ecosystem entities underlying the supply of input ES and their role in the elaboration of agricultural production. A transversal analysis of these ES makes it possible to identify four major types of biophysical determinants for the level of ES supply:

- the nature and spatiotemporal distribution of planned biodiversity at the level of the field, as determined by the selection of species and genotypes by the farmer (sown, planted, grazed), including the intensity and frequency of grazing of grasslands by livestock, where applicable, which will determine the nature of the plant cover and some soil characteristics;

- the nature and spatiotemporal distribution of associated plant (weed species) and animal biodiversity at the field level, resulting from interactions between ecological processes as determined by planned biodiversity, landscape composition and configuration, and factors external to the ecosystem (climate, external agricultural practices);

- soil characteristics (particularly organic matter levels) and abiotic properties (particularly bulk density and available water reserve) resulting from interactions with planned and associated biodiversity, which will affect soil fertility;

- the composition and configuration of the landscape surrounding the agricultural ecosystem, which influence material and energy flows at the supra-field level as well as the movements of some components of the associated biodiversity.

Figure 7-1. Schematic representation of the major biophysical determinants and external factors involved in primary agricultural production

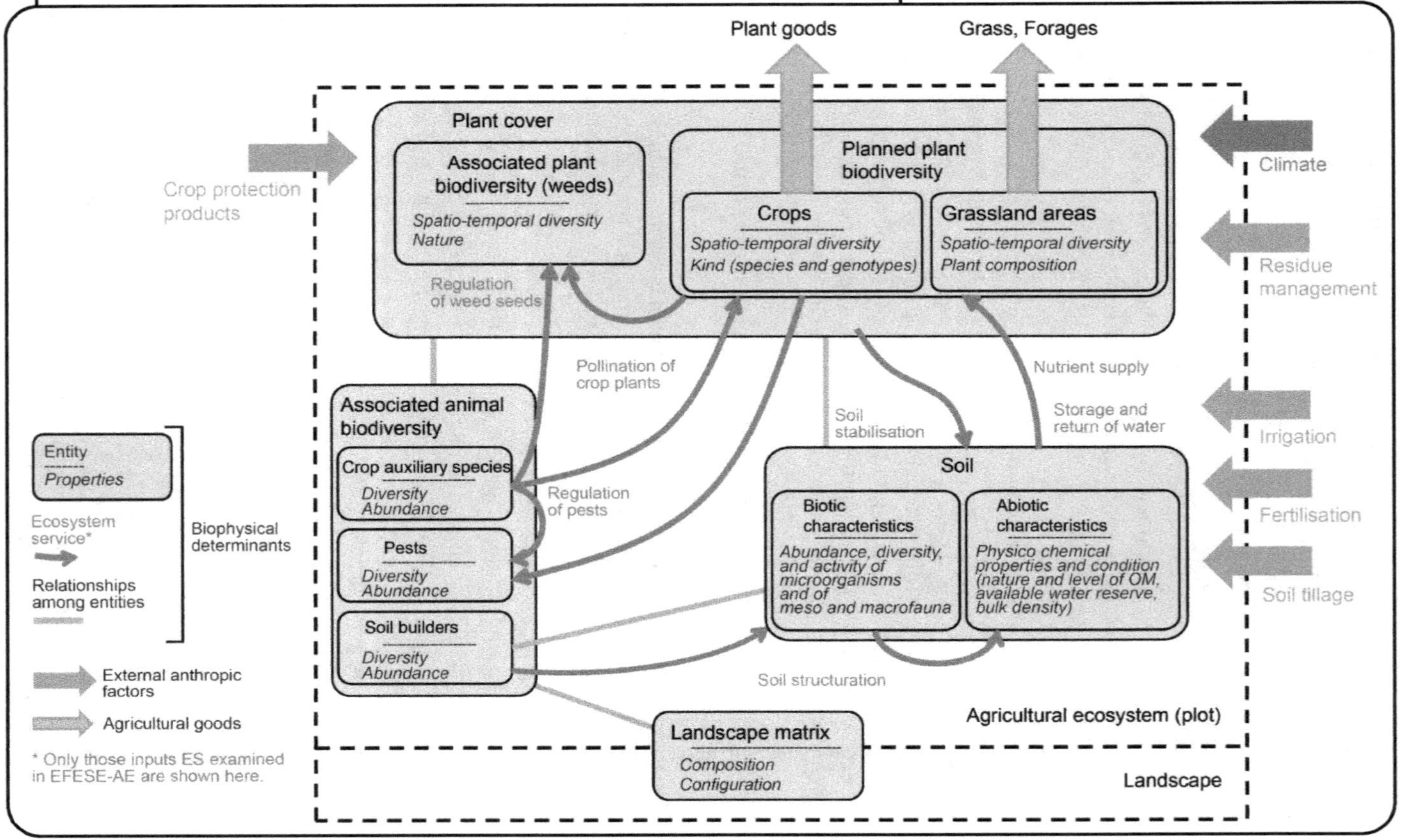

Two types of effects of external practices on the agricultural ecosystem may be distinguished: i) the modification of abiotic soil characteristics, which influence ES relative to the C, N, and water cycles, and ii) the disruption, whether positive (e.g., organic fertilisation) or negative (soil tillage, crop protection products), of associated biodiversity, both plant and animal, which will influence the level of ES reliant on these species.

▮ Quantifying the absolute level of supply of input ES

Quantification of the ES "supply of mineral N to crop plants," "storage and return of water to crop plants," and "soil stabilisation and erosion control" highlights the importance of crop rotations in determining the level of supply of these ES, especially the type of crop (spring vs. winter crops) involved in the rotation and the use or non-use of cover crops. Climate also plays a key role, of course. Selected indicators make it possible to estimate the absolute level of these ES and to observe their spatial distribution, but not to qualify ranges of values in terms of "low" or "high" levels of ES. To effectively inform decision-making, it would be necessary to assess to what degree the ecosystem supplies crop requirements, particularly for mineral N and water, within a given pedoclimatic context. In other words, the indicators must be reconfigured to determine the level of supply of these ES relative to agricultural production requirements.

The level of supply of the "pollination of crop plants" ES is primarily determined by landscape composition and configuration and by climate. Although pollination is one of the most widely studied ES in the literature, to date there has been no indicator allowing direct estimation of its level of supply. However, a newly developed indicator designed to estimate the effects of pollination on agricultural production suggests that this is a limiting factor for pollination-dependent crop yields in multiple regions of France.

In the absence of data that would make it possible to assess the level of supply of the "regulation of pest species" services for France as a whole (e.g., measurements of predation or parasitism percentages, yield losses), only a few estimates of the potential of these ES, based on international and local data, were explored in EFESE-AE. Preliminary results, extrapolated to the whole of France, suggest trends in the spatial variation of levels of regulation of weed seeds by granivorous beetles and the regulation of aphids, but these relationships have only been validated for very specific pedoclimatic and agronomic contexts.

Further work is needed to validate the methods used and preliminary assessment results. The consideration of cropping system characteristics – including tillage intensity, type and degree of soil cover, type and intensity of crop protection applications, type and intensity of fertilisation – represents an important area for future research that could potentially lead to a profound rethinking of the relative significance of cropping system vs. landscape factors in the determination of these ES.

▌ An initial step toward quantifying the part of agricultural production attributable to the input ecosystem services

On average, at the level of crop rotations as currently practiced in France, the portion of crop production attributable to the input ES "N and water" is approximately 50%. Examined on a crop-by-crop basis, this portion varies considerably. A crosswise analysis of the results presented in Chapter 3 enables us to identify major trends in the relationship between the input ES "N and water," fertilisation and irrigation practices, and levels of agricultural production.

First, situations with low average levels of the ES "supply of N to crop plants" and "storage and return of water to crop plants" do not necessarily correspond to situations in which the part of production attributable to these input ES is low. For example, wheat monoculture located in the southern half of the Rhône basin corresponds to situations with low levels of the input ES "N and water," but also to situations where the part of production attributable to these two ES is relatively high. The level of each ES must be examined with regard to crop requirements within the rotation and in the specific pedoclimate.

At the level of the rotation, low production percentages attributable to the input ES "N and water" are strongly correlated with the presence of (grain) maize monoculture in climatic zones characterised by dry summers, or on soils characterised by a low available water reserve, and thus low levels of the ES "storage and return of water to crop plants." In these situations, irrigation appears to be indispensable to maintaining high yield levels. Further description of situations in which irrigation accounts for the overwhelming majority of crop water requirements (more than 75%) would require a more detailed analysis.

Wheat, which has a production percentage attributable to the input ES "N and water" that is relatively high on average, has a tendency to pull the average results obtained at the scale of the rotation upward, given its frequency in the simulated crop sequences. This phenomenon is magnified where wheat is in rotation with sunflower (e.g., on the clayey-limestone plateaus of the Southwest), one of the rare major field crops managed with few inputs (fertilisers and pesticides), and thus has the highest production percentages attributable to the input ES "N and water." Conversely, in crop sequences such as wheat-(wheat-)oilseed rape (or wheat-barley-oilseed rape), production percentages attributable to the input ES "N and water" are pulled slightly downwards by the presence of oilseed rape. Oilseed rape has high N requirements and is primarily grown in the greater Parisian basin where average levels of ES for the supply of mineral N are low. As a result, the average production percentage for oilseed rape that is attributable to the input ES "N and water" is low.

Note that at the level of the cropping system, the total quantity of mineral N available – the sum of the quantity of mineral N available at sowing, supplied by the ecosystem during crop growth, and supplied by fertilisation – is greater than crop needs in all of the situations analysed. These results suggest there is significant room for improvement in converting ecosystems' capacity to supply N (i.e., potential level of ES) into an effective N supply for crop plants (i.e., effective level of ES), and thus enabling a significant reduction in external N inputs (a benefit derived from the ES by the farmer). The need here, above all, is to develop fertilisation management tools and application technologies for nitrogenous fertilisers that allow for input optimisation to meet crop needs, taking into account both the dynamics of crop uptake and the dynamics of ES for supply of mineral N.

Ecosystem services provided by agricultural ecosystems to society as a whole

In addition to input ES, agricultural ecosystems contribute to the supply of other ES, from which society as a whole benefits (including farmers as members of society). These ES are essentially of two types:

- regulating ES that assist in moderating phenomena that are detrimental to human well-being, such as climate change or the movement of pollutants through different environmental compartments;
- so-called "cultural" ES, which provide society with recreational, aesthetic, and/ or spiritual benefits.

■ The complementarity of "ecosystem services" and "impacts" approaches

Agricultural ecosystems have an impact on the biophysical quality of human life through their contribution to the ES of "regulation of water quality" (for a variety of uses) and "global climate regulation". A transversal analysis of the characteristics of these ES suggests that three ecosystem elements play a central role in determining levels of ES supply: i) the spatiotemporal configuration of the ecosystem (managed plant cover, animals), ii) soil organic matter, and iii) soil biodiversity.

Note that although these ES are provided by biophysical determinants at the agricultural field level, the geographical area within which beneficiaries receive ES benefits is much larger. Thus, ES regulating water quality are potentially expressed at the scale of the watershed, while ES for global climate regulation is expressed at the scale of the entire planet.

In addition to assessing the levels of these two ES, the level of the negative impacts on the environment resulting from agricultural activities was estimated. Analysis of the results obtained from these two types of assessments, ES and environmental impacts, suggests that they provide complementary information about agricultural ecosystems: information as the level of one does not make it possible to directly infer the level of the other.

With respect to water quality regulation, the level of ES is higher as the amount of biomass produced is higher: it is above all the process of N retention in harvested biomass or biomass returned to the soil that determines the level of the ES. Also, as has been shown in previous works, soil cover during periods of potential runoff (P regulation) or drainage (N and DOC regulation) is a major determinant of the level of this ES. The evaluation also shows the interest of not simply estimating an absolute level of SE delivered: in certain situations, a small quantity of non-leached N (indicator of the absolute level of SE supply) corresponds to a very large share of N entering the system (fertiliser inputs, mineralisation). This is the case in sunflower plots, where fertiliser inputs are low.

The impact of current cropping systems on water quality was estimated using two indicators: the quantity of leached N and the nitrate (NO_3^-) concentration in drainage water. These two impact indicators are negatively correlated with the indicators of absolute and relative supply of the ES, suggesting that the impact of the cropping system on drainage water quality tends to be lower when the ES level is high. The correlation is not very strong, however, reflecting the fact that a given capacity of N "retention" by the "soil-plant" system can be associated with very different impact levels. For example, situations in which 80% of the N entering a system is not leached can be associated with quantities of leached N ranging from 20 to 100 kg N/ha/yr, and with NO_3^- concentrations varying from less than 50 mg NO_3^-/l to more than 150 mg NO_3^-/l.

With respect to global climate regulation, the two key ES components, conservation of C stocks and additional C storage, are strongly linked: for a given set of agricultural practices and a given pedoclimate, the current level of stored C strongly determines the level of potential additional storage. This interaction is at the root of a potentially troubling conclusion: it is in areas with the lowest levels of stored C – in other words, in intensive major field crop areas – that one can primarily find ecosystems that allow for additional C storage, whereas in livestock areas, the agricultural systems in place may lead to a future loss of C storage. Values obtained for systems that accumulate C are generally below 0.2% and rarely higher than 0.3%; in other words, well below the annual increase of 0.4% of stored soil C targeted by the international "4 per 1000" initiative.

The impact of current cropping systems on climate was assessed in terms of the net annual movement of CO_2 and N_2O between the agricultural ecosystem and the atmosphere (flows of CH_4 being considered as negligible for major field crops). The results show that the vast majority of cultivated agroecosystems are sources of GHG (net GHG emitters), primarily as a result of N_2O emissions. Only a very few simulated cases are sinks for GHG as a result of their low levels of N_2O emissions. Emissions of N_2O increase as external N inputs increase, confirming the significant role of nitrogenous fertilisation in these emissions. The relationship is nevertheless variable as a result of the many factors involved in N_2O production (temperature, water levels, pH, etc.). Finally, the use of cover crops has a favorable effect on GHG budgets. Cover crops increase soil C storage and reduce N_2O emissions. A preliminary qualitative comparison of the spatial distribution of these results again suggests that a single level of net GHG footprint may be linked to different levels of ES and *vice versa*.

We should note that in the analysis presented above there is no explicit distinction made between the status of processes of C storage and processes of C release by the agricultural ecosystem. In reality, however, whereas processes of C storage correspond to one of the two components of the ES, C release is properly understood as a dis-service or a negative impact of agricultural activities (see Chapter 1). Clarification of the status of processes of C release from agricultural ecosystems is thus needed.

▌ The need for a multi-ecosystem approach to characterise cultural services

Agricultural ecosystems form landscapes that people often consider as attractive for the pursuit of recreational activities. As defined within the CICES typology, recreational services correspond more closely to a typology of landscape uses and/or values than to ecosystem services in the sense adopted in EFESE-AE. As with ES of biological regulation, recreational potential is a product of the total landscape. In contrast, however, recreational potential is also expressed at the scale of the total landscape, a space within which several types of ecosystems exist side by side. An approach focusing on major ecosystem type thus appears less relevant for analysing recreational services, since these are provided both by nature and configuration of different ecosystem types within the landscape. The elements presented in EFESE-AE thus seek to propose some alternative avenues for redefining recreational services and to identify the limits associated with existing assessment methods.

From biophysical assessments to economic assessments of ecosystem services

The application of existing economic valuation methods to the ES examined in EFESE-AE illustrates the importance of correctly characterising the biophysical mechanisms underlying ES supply, and identifying suitable indicators for their quantification, prior to applying economic theories and approaches to the question of ES. Indeed, even where these conditions are met, results obtained from economic assessments must be treated with caution, giving due recognition to the assumptions inherent in the calculations and the other limitations belong to the different approaches.

The replacement costs method, which uses the implementation cost of substitution technologies and has been applied to the ES for supply of mineral N and storage and return of water to crop plants, must only be applied at a fine level of spatial resolution. The values obtained for these two ES cannot be summed, although estimated using the same methodology, because these two evaluations are based on different reference situations and substitution technologies. To sum these two values would amount to assuming that the production factors involved are fully substitutable, which is not the case, notably due to the biophysical interactions among processes relating to N and water.

The avoided damages method, which involves estimating the part of production attributable to an ES, is strongly linked to the quality of the underlying biophysical assessment. Updating the biophysical assessment of the ES for crop pollination based on this method offers a good illustration of these challenges. Although both types of economic assessment produce a monetary value, the differences between the two methodologies mean that the two series of results should not be directly compared.

In cases where the biophysical characterisation of the level of ES supply is based on indirect indicators, the methods cited above are not applicable. To overcome these obstacles and make use of the avoided damages method, additional research is needed on the quantitative relationships between ES, production outputs, agricultural practices, and landscape management.

 Towards the management of ecosystem services supply

The analysis of ES bundles, widely advanced as a decision-making tool, can assist in characterising the range of ES supplied across a given territory (levels of supply for a specific group of ES). It continues to present methodological challenges, and its implementation at scales relevant to the issues at stake in territorial policymaking, covering a mosaic of ecosystems, is not fully operational. In addition, the ES bundles approach provides a snapshot of ES levels a moment in time t, without any information as to the nature of interactions among ES. A detailed understanding of biophysical interactions among ES, obtained through a sophisticated biophysical description of each ES in turn, is required in order to predict the effects of different actions on variations in levels of ES supply.

Research conducted in EFESE-AE has provided an initial level of information on the key biophysical determinants of ES supplied by agricultural ecosystems, and the major action levers to be considered. Subsequent research is nonetheless needed to pursue this analysis. In the short term, by drawing on the disciplinary expertise mobilised here, it should be possible to enrich the descriptions by supplying the direction of the correlations (positive or negative) existing among the different components, and the relative importance of the different components. It would then be possible to construct "fuzzy cognitive maps" of these interactions and, in a subsequent step, to infer the effects of changes in levels of external factors on levels of ES. Eventually, this type of representation would also allow for the consideration of feedback loops. In the medium term, this type of analysis should allow for an improved understanding of the antagonisms, synergies and convergent effects that exist among different ES depending on different agricultural ecosystem characteristics and, more broadly, different characteristics of agricultural landscapes. Similarly, the use of dynamic simulation models of the soil-plant-animal system offer the possibility of conducting in-depth analyses of the interactions among simulated ES as a function of different production contexts.

Finally, the levers identified above will not be the same for other types of ES, ecosystems, and/or beneficiaries. Multi-criteria assessment methods are needed, in which the environment is represented using indicators for three key sub-domains corresponding to the principal management issues associated with agricultural ecosystems: i) levels of ES provided to the different beneficiaries under consideration; ii) the environmental impacts of agricultural activities; iii) the conservation of biodiversity (not including cultural heritage). Accounting for dis-services, not addressed in EFESE-AE, will also be necessary. These types of methods should make it possible to identify antagonisms and synergies within each sub-domain (e.g. among ES provided to farmers or to society) or between sub-domains (e.g. between the supply of ES and environmental impacts) at appropriate levels of organisation (field, agroecosystem, landscape, etc.).

8. Perspectives for future research and other developments

Develop more precise sources of data

In the first place, the assessments could be improved by better access to more detailed data on soils and land use (nature of plant cover, soil and biomass management practices).

▌ Soils database

The assessment method developed in EFESE-AE made use of data on soil characteristics obtained by applying pedotransfer functions to the qualitative information contained in the geographical database of French soils (*Base de Données Géographique des Sols de France*) at a scale of 1:1,000,000. The uncertainty associated with these functions has not been formally evaluated, but is potentially high. The use of the new soil map at a scale of 1:250,000 (accessible since 2020) – which contains quantified information on soil type and some soil characteristics, notably clay content and soil organic matter levels – would allow for a more precise assessment of soil properties.

Particular attention should also be paid to the determination of two key soil characteristics/properties, organic matter levels and bulk density. In the current method, these are determined without any distinction made between cropping systems vs. grassland systems. For example, the estimation of organic matter levels for temporary grasslands made in Mulder *et al.* (2015[58], 2016[59]) makes no effort to account for the length of time in grass for these soils. In the French soils database, bulk density is reported as being identical for major field crops and for grasslands. Information currently available on the spatial distribution of plant cover sequences (rotations) offers good potential for improvement of these data. Indeed, it should be possible to combine data on organic matter levels and bulk density from the French Soil Quality Measurement Network (*Réseau de Mesures de la Qualité des*

---

58. Mulder VL, Lacoste M, Martin MP, *et al.* (2015) Understanding large-extent controls of soil organic carbon storage in relation to soil depth and soil-landscape systems. Global Biogeochem Cycles 29:1210–1229. doi: 10.1002/2015GB005178
59. Mulder VL, Lacoste M, Richer-de-Forges AC, Martin MP, Arrouays D (2016) National versus global modelling the 3D distribution of soil organic carbon in mainland France. Geoderma 263: 16-34. doi: 10.1016/j.geoderma.2015.08.035

Sols – RMQS), on the one hand, and data on crop rotations, on the other, in order to improve the estimate of these two soil properties as a function of plant cover sequences present in the field.

❚ Soil and biomass management practices

The representation of the spatial and temporal distribution of agricultural practices for different crops constitutes a major limitation of the simulation method as applied to cropping systems. The only large-scale information currently comes from the "Agricultural Practices" surveys conducted every 5 years by the French Ministry of Agriculture. The scale of statistical representativeness of this database (the administrative region) constitutes the primary limiting factor to its use for this type of analysis. Access is needed to (i) annual data in order to infer the inter-annual variability of practices as a function of climate and (ii) a finer level of spatial resolution (at least on the order of the SAR), in order to account for the variability of these practices as a function of the diversity of pedoclimatic and agricultural contexts. A working group of the GIS "Major field crops for high economic and environmental performance" (GCHP2E) is currently in the process of developing a strategy for advancing this effort. The goal is to develop a spatialised model of agricultural practices at a detailed level by combining existing databases on agricultural systems and pedoclimates.

❚ Plant covers

The crop rotations represented in EFESE-AE came from an Inra database resulting from the analysis of the annual information from the French LPIS. This database offers numerous opportunities for the identification of prevailing crop rotations as well as alternative crop rotations. However, only a simplified version of the French LPIS is accessible to the scientific community. As a result, the Inra crop rotation database only contains information as to major crop types (e.g., wheat and other cereals). To improve the representation of sown crops (e.g. distinguishing between different cereals), all of the information in the French LPIS should be accessed, full access to the French LPIS data is needed.

The French LPIS does not include information on cover crops, whereas the preliminary results of the assessment have shown that cover crops can influence the supply of ES by cropping systems. Information from the French LPIS should thus be linked with satellite data on the presence and duration of cover crops in the field. Some work currently underway – for example, UMR CESBIO[60] is developing methods to map the presence of cover crops across France – should soon make it possible to supply this type of information at a detailed level of spatial resolution.

---

60. https://www.cesbio.cnrs.fr/en-cesbio/

Move from absolute to relative levels of ecosystem services supply

ES assessments completed in EFESE-AE – that is, the quantification of absolute levels of ES supply – constitute an initial step in the effort to assess the range of ES currently supplied by agricultural ecosystems. For some ES, however, the results of this assessment suggest a need to go beyond the quantification of absolute values and toward the quantification of ES levels relative to specific factors linked to their use or management. Initial proposals were conducted for the ES for the supply of mineral N to crop plants and storage and return of water to crop plants, and for soil stabilisation and erosion control. This effort should be completed for the remaining ES, and should be accompanied by a reflection on the "demand" for the ES or on reference situations that could be used for these "relative" assessments.

Move from the potential level to the effective level of ecosystem services supplied by ecosystems

The "effective" level of ES could not be estimated for all the ES examined here. In some cases, only the ecosystem's capacity to supply the ES (or the "potential" level of the ES) could be quantified. This was particularly true for ES provided to farmers. In some contexts, moreover, the difference between effective and potential ES levels may be significant. Information on effective ES levels, and on the adaptation of agricultural practices in response to variations in ES, is needed in order to provide a robust economic assessment of this type of ES. To move toward the estimation of effective ES levels by ecosystems, it will be necessary (1) to develop more direct indicators of levels of ES, and then (2) develop information on the ways in which beneficiaries modify their behavior (agricultural practices, management methods) when they become aware of the levels of ES provided by the ecosystem.

With regard to the first of these points, due to a lack of data or other necessary research, some ES were not estimated through a direct indicator of the process or state that defines the ES. They were estimated through the quantification of a biophysical determinant of the corresponding process or state. For example, for the ES for the regulation of pest insects, a landscape composition indicator (a biophysical determinant of this ES) was used to predict the potential level of regulation. For the ES for crop pollination and the regulation of weed seeds by carabids, the biophysical assessment was even more indirect: indicators of landscape composition/configuration were used to predict the abundance of crop auxiliary species, which is a biophysical determinant of these ES. But correlations between landscape characteristics and levels of regulation are often somewhat loose, and thus only provide a very indirect and often imprecise estimate of the ES. To estimate the level of ES, it would thus be necessary to develop more direct indicators, in other words, for the level of pest regulation.

With regard to the second point, some ES require the addition of human or material capital for the benefit to be obtained. Information is thus needed on the practices of beneficiaries if one is to move from the quantification of potential ES levels to effective ES levels. The practices of the ecosystem manager can have effects on potential ES levels. For example, methods for estimating the ES for the storage and return of water to crop plants, and the way in which the level of this ES is accounted for in deciding on an irrigation strategy, will determine the precise effective level of this ES. For ES for natural biological control, interactions between potential ES levels and agricultural practices need to be determined. This would make it possible to estimate the effective level of ES, in other words, the level of natural biological control the farmer benefits considering both his or her practices (e.g., pesticide applications) and the potential level of biological pest control. Estimating yield losses (e.g., negative impacts on production quality or quantity) avoided thanks to ES for biological control is key both to determining the relationship between ES levels and agricultural practices and to the completion of an economic assessment of this ES. Few models are available for assessing the links between ES, agricultural practices, and yield damages, however. This is thus another area for future research.

 Explore other types of cropping systems

Designing production systems based more on ES and less on external inputs will require advanced research on the interactions between ecosystem configuration, landscape configuration, external agricultural practices, climate, ES levels and levels of production of agricultural goods (notably to identify negative or positive correlations). Initial studies of processes of biological regulation suggest the potential for a profound rethinking of the relative importance of landscape configuration and composition relative to cropping system effects, at least for cropping systems with certain characteristics (e.g., permanent cover with little soil disturbance).

In the short term, the comparative analysis of various types of low-input cropping systems and/or production systems – for example, conservation agriculture, integrated crop-livestock systems – should make it possible to advance in this direction. The development of models to simulate the effects of a wide range of ecosystem configurations and external agricultural practices on different ES levels could assist in the development of strategies for agricultural ecosystem management that avoid or minimise potential antagonisms among ES.

Further work to be done on the status and role of livestock in the supply of ecosystem services

In EFESE-AE, agricultural livestock present in the ecosystem (that is, not reared indoors) are considered to be biotic components of the ecosystem. They belong to the category of planned biodiversity. By analogy with plant biodiversity, agricultural practices that determine the spatiotemporal distribution of animals within the ecosystem are described as ecosystem configuration practices. Agricultural livestock on pasture are also a means of production of agricultural goods.

This preliminary understanding of the status and role of livestock animals in the ecosystem from the point of view of ES remains to be further developed. EFESE-AE sought to quantify the level of production of agricultural goods made possible by crop production within the same territory. In future work, it will be also be important to assess the role of livestock as organisms involved in the provision of other ES (for example, "regulation of animal livestock diseases").

In addition, further research is needed to assess the percentage of livestock production attributable to input ES. In the short term, a first step would be to combine the approach adopted to quantify the share of plant production attributable to the functioning of the ecosystem, with the quantification of the level of production of animal goods enabled by plant production in the local territory. Secondly, it would be necessary to develop a more integrative approach to the analysis of the soil-plant-animal system.

Consider the resilience of ecosystem services to future change

Alterations to ecosystems, whether from the effects of climate change or from changes in territorial management (e.g., urbanisation, reforestation), will necessarily have an impact on ES supply. Identifying the conditions for resilience for ES supply in the face of these different types of change will require identifying the key biophysical and socio-economic properties supporting such resilience. Potential strategies to maintain ES levels in the face of these changes, or to shift ES so as to better align with human priorities, could thus be determined.

Most studies in ecology are focused on temporal stability, often defined as the reverse of temporal variability for a measured value (for example, productivity). Such studies have often demonstrated a positive correlation between biodiversity and ecosystem temporal stability. However, recent review articles, as well as expert opinion, on the relationship between ecosystem properties and the resilience of ES suggest that in addition to species diversity and functional diversity, ecological connectivity between ecosystems and the condition of slow-moving variables (e.g., soil organic matter levels) have a strong impact on the resilience of ES provided by these systems. Here again, the determination of optimum levels for each of these three key properties, how they interact, and how they relate to different pedoclimatic contexts, or even landscape contexts, constitute critical areas for future research.

From a methodological point of view, the development of dynamic models for soil-plant-animal systems and landscapes should, in time, offer the possibility of analysing the spatiotemporal dynamics of ES and ES interactions, and thus a means of understanding the variability, and potentially the resilience, of ES to climate change and other anthropic factors.

Appendix 1. Composition of the working group[61]

 **Methodological and scientific coordination**

Olivier Therond, Inra, UMR1132-LAE « Laboratoire agronomie et environnement », Scientific co-lead.

Anaïs Tibi, Inra, UAR1241-DEPE « Délégation à l'expertise scientifique collective, à la prospective et aux études », Project coordination.

Muriel Tichit, Inra, UMR148-SADAPT « Sciences pour l'action et le développement : activités, produits, territoires », Scientific co-lead.

 **Transversal engineering and data integration in an "ecosystem services" information system**

Éric Cahuzac (responsible), Inra, US0685-ODR « Observatoire des programmes communautaires de développement rural ».

Annette Girardin, Inra, UAR1241-DEPE « Délégation à l'expertise scientifique collective, à la prospective et aux études ».

Anne Meillet, Inra, US0685-ODR « Observatoire des programmes communautaires de développement rural ».

Thomas Poméon, Inra, US0685-ODR « Observatoire des programmes communautaires de développement rural ».

 **Coordinating experts**

In italics, theme for which the expert coordinated the analysis.

* : experts who have also participated in data engineering.

David Bohan, Inra, UMR1347 « Agroécologie » : *Regulation of weed seeds.*

Philippe Choler, CNRS, UMR5553-LEA « Laboratoire d'écologie alpine » : *Soil stabilisation and erosion control.*

---

61. affiliations of people at the time of the project.

Julie Constantin*, Inra, UMR1248-AGIR « Agroécologie, innovations, territoires » : *Regulation of water quality.*

Isabelle Cousin, Inra, UR0272-SOLS « Science du sol » : *Supply of mineral N to crop plants, Storage and return of water.*

Maia David, AgroParisTech, UMR0210-ECO-PUB « Économie publique » : *Economic assessment of ecosystem services.*

Philippe Delacote, Inra, UMR0356-LEF « Laboratoire d'économie forestière » : *Economic assessment of ecosystem services.*

Michel Duru, Inra, UMR1248-AGIR « Agroécologie, innovations, territoires » : *Agricultural plant goods.*

Magali Jouven, Montpellier SupAgro, UMR0868 SELMET « Systèmes d'élevage méditerranéens et tropicaux » : *Agricultural animal goods.*

Yves Le Bissonnais, Inra, UMR1221-LISAH « Laboratoire d'étude des interactions sol-agrosystème-hydrosystème » : *Soil stabilisation and erosion control.*

Fabrice Martin-Laurent, Inra, UMR1347 « Agroécologie » : *Natural attenuation of pesticides by soils.*

Vincent Martinet, Inra, UMR0210-ECO-PUB « Économie publique » : *Economic assessment of ecosystem services.*

Maud Mouchet*, MNHN, UMR7204-CESCO « Centre d'écologie et des sciences de la conservation » : *Service bundles analysis.*

Sylvain Pellerin, Inra, UMR1391-ISPA « Interaction sol-plante-atmosphère » : *Global climate regulation.*

Sylvain Plantureux, université de Lorraine, UMR1132-LAE « Laboratoire agronomie et environnement » : *Agricultural plant goods.*

Emmanuelle Porcher, MNHN, UMR7204-CESCO « Centre d'écologie et des sciences de la conservation » : *Pollination of crop plants.*

Laurence Puillet*, Inra, UMR791-MoSAR « Modélisation systémique appliquée aux ruminants » : *Agricultural animal goods.*

Tina Rambonilaza, Irstea « Environnement, territoire, infrastructure » : *Economic assessment of ecosystem services.*

Bénédicte Rulleau, université de Versailles St Quentin, Irstea, EA4455-CEARC « Cultures, environnements, Arctique, représentations, climat » : *Economic assessment of ecosystem services.*

Adrien Rusch, Inra, UMR-1065-SAVE « Santé et agroécologie du vignoble » : *Regulation of insect pests.*

Jean-Michel Salles, CNRS, UMR1135-LAMETA « Laboratoire montpelliérain d'économie théorique et appliquée » : *Economic assessment of ecosystem services.*

Léa Tardieu, Inra, UMR0356-LEF « Laboratoire d'économie forestière » : *Economic assessment of ecosystem services.*

 ## Scientific contributors

Members of the working group who contributed to the drafting or to the discussions that determined the way in which the questions were approached or structured.

* : contributors who have also participated in data engineering.

Francesco Accatino*, Inra, UMR148-SADAPT « Sciences pour l'action et le développement : activités, produits, territoires ».

Christian Bockstaller, Inra, UMR1132-LAE « Laboratoire agronomie et environnement ».

Thierry Bonaudo, AgroParisTech, UMR148-SADAPT « Sciences pour l'action et le développement : activités, produits, territoires ».

Maryline Boval*, Inra, UMR791-MoSAR « Modélisation systémique appliquée aux ruminants ».

Bruno Chauvel, Inra, UMR1347 « Agroécologie ».

Maguy Eugène*, Inra, UMR1213 UMRH « Unité mixte de recherche sur les herbivores ».

Colin Fontaine*, CNRS, UMR7204 CESCO « Centre d'écologie et des sciences de la conservation ».

Ilse Geijzendorffer, Tour du Valat.

Anne-Isabelle Graux*, Inra, UMR1348 PEGASE « Physiologie, environnement et génétique pour l'animal et les systèmes d'élevage ».

Barbara Langlois, Inra, UMR0210 ECO-PUB « Économie publique ».

Robert Lifran, retraité, ex-Inra.

Gabrielle Martin*, CNRS, UMR7204 CESCO « Centre d'écologie et des sciences de la conservation ».

Orla McLaughlin, Inra, UMR1347 « Agroécologie ».

Catherine Mignolet, Inra, UR0055 ASTER « Agro-systèmes territoires ressources Mirecourt ».

Marie-Odile Nozières-Petit, Inra, UMR0868 SELMET « Systèmes d'élevage méditerranéens et tropicaux ».

Ole P. Ostermann, Commission européenne, Joint Research Centre (JRC).

Maria Luisa Paracchini, Commission européenne, Joint Research Centre (JRC).

Sandrine Petit-Michaut, Inra, UMR1347 « Agroécologie ».

Jean-Louis Peyraud, Inra, UMR1348 PEGASE « Physiologie, environnement et génétique pour l'animal et les systèmes d'élevage ».

Thomas Poméon*, Inra, US0685 ODR « Observatoire des programmes communautaires de développement rural ».

Françoise Ruget, Inra, UMR1114 EMMAH « Environnement méditerranéen et modélisation des agro-hydrosystèmes ».

Daniel Sauvant*, Inra, UMR791 MoSAR « Modélisation systémique appliquée aux ruminants ».

Céline Schott*, Inra, UR0055 ASTER « Agro-systèmes territoires ressources Mirecourt ».

 ## Data engineering

Membres du groupe de travail qui ont participé à la collecte, la gestion et le traitement des données, en appui aux experts et aux contributeurs scientifiques.

Designing the STICS/PaSim simulation system

Éric Casellas, Inra, UR0875 MIAT « Mathématiques et informatique appliquées Toulouse ».

Laetitia De Sousa, ex-Inra, UAR1241 DEPE « Délégation à l'expertise scientifique collective, à la prospective et aux études ».

Christine Le Bas, Inra, US1106 « INFOSOL ».

Raphaël Martin, Inra, UMR0874 UREP « Unité mixte de recherche sur l'écosystème prairial ».

Hélène Raynal, Inra, UR0875 MIAT « Mathématiques et informatique appliquées Toulouse ».

Rémi Resmond, Inra, UMR1348 PEGASE « Physiologie, environnement et génétique pour l'animal et les systèmes d'élevage ».

Dominique Ripoche, Inra, US1116 « AGROCLIM ».

Evaluation of agricultural goods

Camille Dross, Inra, UMR148 SADAPT « Sciences pour l'action et le développement : activités, produits, territoires ».

Benoît Garcia, Inra, US0685 ODR « Observatoire des programmes communautaires de développement rural ».

Élise Maigné, Inra, US0685 ODR « Observatoire des programmes communautaires de développement rural ».

Calypso Picaud, Inra, UR0055 ASTER « Agro-systèmes territoires ressources Mirecourt ».

Thomas Puech, Inra, UR0055 ASTER « Agro-systèmes territoires ressources Mirecourt ».

Joao Pedro Domingues Santos, Inra, UMR148 SADAPT « Sciences pour l'action et le développement : activités, produits, territoires ».

▌ Evaluation of the ecosystem services "Weed seed regulation" and "Regulation of insect pests"

Luc Biju-Duval, Inra, UMR1347 « Agroécologie ».

Stéphane Derocles, Inra, UMR1347 « Agroécologie ».

▌ Assessment of the "Soil stabilisation and erosion control" ecosystem service

Joël Daroussin, Inra, UR0272 SOLS « Science du sol ».

▌ Documentation

Virginie Lelièvre, Inra, « Département environnement et agronomie ».

 DEPE project team

Marc-Antoine Caillaud, support for the organisation of the conference.

Kim Girard, logistics and financial management.

Annette Girardin, project management support.

Anaïs Tibi, project management, writing the condensed report and the summary, editorial coordination.

Appendix 2. Correspondence between the CICES v4.3 typology and the list of ecosystem services examined in EFESE-AE

CICES typology	EFESE-AE typology
Provisioning services	**Production of agricultural goods**
Cultivated crops Wild plants, algae and their outputs	Production of plant goods from cultivated plants
	Fodder production from grassland
	Production of wild plants for purposes other than fodder
Plants and algae from in situ aquaculture	*Not studied**
Reared animals and their outputs	Production of animal goods
Wild animals and their outputs	*Not studied*
Animals from in situ aquaculture	*Not studied*
Surface water for drinking	Storage and return of water (regulating service)
Ground water for drinking	
Fibers and other materials from plants, algae and animals for direct use or processing	*Not studied*
Materials from plants, algae and animals for agricultural use	Plant biomass: Production of plant goods from cultivated plants
	Animal biomass: Not studied
Genetic materials from all biota	*Not studied*
Surface water for non-drinking purposes	Storage and return of water (regulating service)
Ground water for non-drinking purposes	
Plant-based resources [for energetic use]	Production of plant goods from cultivated plants
Animal-based ressources [for energetic use]	*Not studied*
Animal-based energy	*Not studied*

CICES typology	EFESE-AE typology
Regulating services	
Bio-remediation by micro-organisms, algae, plants, and animals	Natural attenuation of pesticides by soils Regulation of water quality with respect to N, P, and DOC *Regulation of air quality: not studied*
Filtration/sequestration/storage/accumulation by micro-organisms, algae, plants, and animals	
Filtration/sequestration/storage/accumulation by ecosystems	
Dilution by atmosphere, freshwater and marine ecosystems	
Mediation of smell/noise/visual impacts	*Not studied*
Mass stabilisation and control of erosion rates	Soil stabilisation and erosion control
Buffering and attenuation of mass flows	
Hydrological cycle and water flow maintenance	Storage and return of water
Flood protection	*Not studied*
Storm protection	*Not studied*
Ventilation and transpiration	*Not studied*
Pollination and seed dispersal	Pollination of crop plants
Maintaining nursery populations and habitats	*Not conceptualised as an ecosystem service*
Pest control	Regulation of weed seeds
	Regulation of insect pests
Disease control	*Not studied*
Group "Soil formation and composition" (Weathering processes - Decomposition and fixing processes)	Supply of mineral N to crop plants Supply of other nutrients to crop plants Soil structuration
Chemical condition of freshwaters	Regulation of water quality with respect to N, P, and DOC
Chemical condition of salt waters	*Not studied*
Global climate regulation by reduction of greenhouse gas concentrations	Global climate regulation by GHG attenuation and C sequestration
Micro and regional climate regulation	*Not studied*

CICES typology	EFESE-AE typology
Cultural services	
Spiritual and/or emblematic interactions	*Not studied*
Other cultural outputs	*Not studied*
Physical and experiential interactions	Recreational potential (outdoor activities, no sampling)
	Recreational potential (outdoor activities, with sampling)
Intellectual and representative interactions	*Not studied*

* The reasons are detailed in the study's extended scientific report.

Appendix 3. Summary of biophysical assessment methods (indicators) used in EFESE-AE to quantify ecosystem services

Ecosystem service	Biophysical indicator(s)	Types of agricultural ecosystems involved in the assessment	- Indicator units - Spatial resolution of the calculation	Nature of the data and tools employed	Assessment robustness [a]
Storage and return of water to crop plants	Amount of water transpired by the commercial crop during its period of growth → *indicator of the effective level of ES*	Major field crops (8 crops)*	- mm of water/yr (average 1984-2013) - PCU [b]	STICS* model	+++
Storage and return of blue water	Water yield → *indicator of the potential ES level*	Major field crops (8 crops)*	- mm of water/yr (average 1984-2013) - PCU	STICS* model	+++
Supply of mineral nitrogen to crop plants	Amount of N supplied by the ecosystem during the cropping cycle (mineralisation + fixation) → *indicator of the potential level of ES*	Major field crops (8 crops)*	- kg of N /ha/yr (average 1984-2013) - PCU	STICS* model	++
Soil stabilisation and erosion control	Difference in erosion rates between the current situation and the reference situation (bare soil) → *indicator of the effective level of ES*	All agricultural land divided into 4 major agricultural ecosystem types [d]	- Metric tons of soil/ha/an (moyenne 2010-11-12) - 100 m square cells	MESALES model	+++
Pollination of crop species	(i) Pollination potential as developed by the MAES program → *indicator of the potential level of ES*	All agricultural land [d]	- No units ("current state" [c]) - 100 m square cells	(i) already calculated by the JRC	++++
	(ii) Indicator of ES based on crop yields → *indicator of the effective level of ES*	Major field crops, vegetable crops, tree crops (58 different crops)	- No units (average 2000-2010) - Department	(ii) analysis of agricultural statistics	+
	(iii) Pollinator species richness → *indicator of the potential level of ES*	Arable land, mixed agricultural areas, grasslands and open agricultural areas [d]	- Number of species-types ("current state") - 100 m square cells	(iii) spatial extrapolation of SpiPOLL data	++

Ecosystem service	Biophysical indicator(s)	Types of agricultural ecosystems involved in the assessment	- Indicator units - Spatial resolution of the calculation	Nature of the data and tools employed	Assessment robustness [a]
Regulation of weed seeds	(i) Potential weed seed abundance → *indicator of the potential level of ES*	Major field crops (cereals, oilseeds), vegetable crops, fallow, temporary grasslands	- Quantity of seeds per m² (average 2010-11-12) - 2 km square cells	(i) analysis of data from the literature	++
	(ii) Potential carabid abundance in wheat fields → *indicator of the potential level of ES*	Winter cereals (wheat and barley)	- Number of individuals (2012) - 2 km square cells	(ii) analysis of data from the literature	+
Regulation of pest insects	Indicator of potential regulation of aphids in wheat, barley, cabbage, and soybean crops → *indicator of the potential level of ES*	Winter cereals (wheat and barley)	- No units (2012) - 2 km square cells	analysis of data from the literature	+
Regulation of water quality with respect to N	Annual average amount of non-leached N → *indicator of the effective ES level*	Major field crops (8 crops)*	- kg N/ha/yr and in % (average 1984-2013) - PCU	STICS* model	+++
Global climate regulation by C sequestration	(i) Quantity of C stored in soil OM and in woody biomass → *indicator of the effective ES level*	All agricultural land [d]	- metric tons of C/ha ("current conditions"[c]) - PCU	(i) Analysis of data from Inra-Infosol and from the literature	++
	(ii) Variation in the annual average stock of C in the soil → *indicator of the effective ES level*	Major field crops (8 crops)*	- % (average 1984-2013) - PCU	(ii) STICS* model	+++

Ecosystem service	Biophysical indicator(s)	Types of agricultural ecosystems involved in the assessment	- Indicator units - Spatial resolution of the calculation	Nature of the data and tools employed	Assessment robustness [a]
Recreational potential (outdoor activities without taking)	Degree of "naturalness" of agricultural ecosystems → *indicator of the potential ES level*	All agricultural land [d]	- no units ("current conditions") - 100 m square cells	Modification of the method used in Paracchini *et al.* (2014) [e]	++

* Simulation plan developed for grassland ecosystems but not implemented in EFESE-AE;

a. Qualitative assessment completed based on expert opinion. The indicator varies from "++++" = robust to "+" = needs additional work to consolidate and verify results prior to further use;

b. PCU = pedoclimatic unit, defined specifically in EFESE-AE;

c. "current state" = the index is not an average of obtained values for a series of years, but is calculated using various data sources from various years;

d. index calculated for all of mainland France, but only those results relating to the "agricultural" pixels are presented in this summary;

e. Paracchini M.L, Zulian G., Kopperoinen L., Maes J., Schägner JP, Termansen M., Zandersen M., Perez-Soba M., Scholefield P.A., Bidoglio G., 2014. Mapping cultural ecosystem services: A framework to assess the potential for outdoor recreation across the EU, Ecological Indicators 45:371-395.

Fichier préparé par Nicolas Perrier, société 4P
Imprimé pour vous par Libri Plureos GmbH (Allemagne)

Editorial coordination: Sylvie Blanchard

Page layout: EliLoCom

Legal deposit on publication: December 2024